CONFLUENCE

CONFLUENCE
A History of Fort Snelling

Hampton Smith

MINNESOTA
HISTORICAL
SOCIETY PRESS

mnhspress.org

The Minnesota Historical Society Press is a member of the Association of University Presses.

Manufactured in the United States of America

10 9 8 7 6 5 4 3 2 1

♾ The paper used in this publication meets the minimum requirements of the American National Standard for Information Sciences—Permanence for Printed Library Materials, ANSI Z39.48-1984.
International Standard Book Number
ISBN: 978-1-68134-156-9 (hardcover)
ISBN: 978-1-68134-157-6 (e-book)

Library of Congress Cataloging-in-Publication Data
Names: Smith, Hampton, 1949- author.
Title: Confluence : a history of Fort Snelling / Hampton Smith.
Other titles: History of Fort Snelling
Description: Saint Paul, MN : Minnesota Historical Society Press, [2021] | Includes bibliographical references and index.
Identifiers: LCCN 2021014878 | ISBN 9781681341569 (hardback) | ISBN 9781681341576 (epub)
Subjects: LCSH: Fort Snelling (Minn.)—History. | Fortification—Minnesota—Fort Snelling—History.
Classification: LCC F614.F7 S65 2021 | DDC 355.709776/57—dc23
LC record available at https://lccn.loc.gov/2021014878

This and other Minnesota Historical Society Press books are available from popular e-book vendors.

CONFLUENCE

Prologue

VISITORS AT FORT SNELLING STATE PARK ENJOY hiking on the trail that circles the shores of Wita Taŋka (Big Island) or Pike Island, where the Minnesota and Mississippi Rivers converge. This place where the rivers meet is known in the Dakota language as Bdote, and for many it is the center of the world, a place of creation, the nexus of a sacred landscape. The confluence and the region around it was and is the homeland of the Dakota—Mni Sota Makoce. Dakota tradition says the island itself served as a meeting place, where large gatherings of people took place every year. Their oyate, or nation, spread from here over a vast area, stretching from what is now Wisconsin to Montana and from Iowa and Nebraska to Manitoba, Saskatchewan, and Alberta. This is also the place where a group of Dakota leaders met with US Army officer Lieutenant Zebulon Pike on September 21, 1805, and agreed to allow the United States to build a military post where the two rivers join. Fifteen years later, the construction of Fort St. Anthony, later known as Fort Snelling, began.[1]

Walking west from the confluence along the Mississippi shore of Wita Taŋka, visitors might catch sight of a bald eagle perched on one of the cottonwood trees along the riverbank, or—at the right time of day—hear the call of a barred owl from the interior woods. There is no visible trace of the centuries of Dakota ceremony, nor of the concentration camp built below the bluff in 1862. But looking upstream, one cannot miss the walls, towers, and gun ports of Fort Snelling, which seem to grow out of the limestone bluff ahead. It dominates the scene, much as it did in the 1820s when it was built. Most hikers do not realize that this imposing structure is something of an illusion. Its walls and buildings are for the most part a careful reconstruction of the fort as it is thought to have appeared shortly after its completion in 1824.

But this is the view from below the bluff. If visitors make the effort to climb the steep path up to the main gate of Historic Fort Snelling, a different scene presents itself. Looking back toward the rivers, they see bridges and highways converging around Bdote and Oheyawahi (Pilot Knob), while to the west, beyond the reconstructed fort, stand the many neat, even stylish brick buildings built by the army in the decades after the original Fort Snelling had ceased to be a colonial outpost.

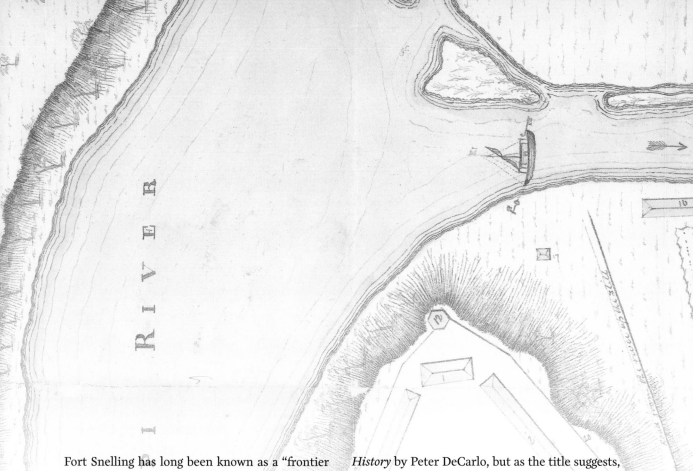

Fort Snelling has long been known as a "frontier fortress." From this vantage point, it becomes apparent that its history is deeper and far more complex.

The story of Fort Snelling has been told from a variety of perspectives, largely concentrating on the early period, 1819 to 1858. Because of the fort's place in the early Anglo-American history of Minnesota, every general history of the state has touched on the creation of the fort in 1819 and 1820 and the historical background of those events. Its place in relation to Indigenous people and the various treaties affecting them is usually discussed as well. These works have little to say about the post's history after the Civil War, however. A number of monographs concentrate on fairly narrow time periods of Fort Snelling's later history, such as the Civil War and World War II. To date, only one complete history of Fort Snelling has appeared, *Fort Snelling at Bdote: A Brief History* by Peter DeCarlo, but as the title suggests, it provides an overview rather than an in-depth history.[2]

This book aims to provide a more expansive history that focuses specifically on Fort Snelling, one that looks into the past before the presence of the United States in the region, then follows the story through the difficult history of colonization and the creation of Minnesota, across the military history of multiple wars, and finally, into the present and our attempts to understand that history. This history can be considered in several phases. The first, from 1819 to 1858, concerns the role of the fort and the Indian agency that was its partner in the European American colonization of the region, concluding with the fort's brief closure and sale in 1858. This complex period receives a thorough treatment, with detailed accounts of life at the fort under Colonel Josiah Snelling, the interactions between the US Indian agent and the Dakota and Ojibwe nations, and the growth of the European American community around the fort.

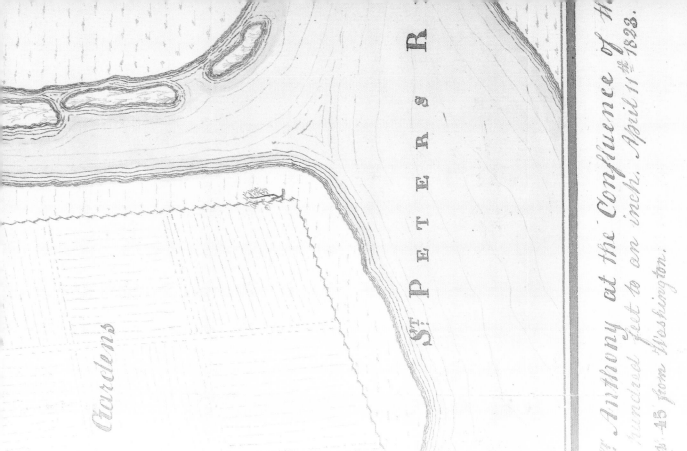

A short but intense period followed from 1861 to 1865, when Fort Snelling was the center of recruitment for Minnesota's Civil War regiments and was also at the heart of the bleak history of the US–Dakota War and subsequent wars on the Great Plains. From 1866 to 1919, the post became a valued part of the US military's infrastructure. It was expanded and developed as a training and support center, culminating in the location of a major military hospital there during and after World War I. From 1920 through 1941, Fort Snelling enjoyed a unique period as the "country club of the army," when the post's leaders, acting as ambassadors for the military, had particularly close ties to the Twin Cities community. This era ended abruptly in 1942, when Fort Snelling became a major induction center for the region's draftees and volunteers during World War II and a training base for specialty units like the army's Military Intelligence Service Language School. The changing nature of warfare and the role of the United States as a world power following World War II saw the end of Fort Snelling as an active military base. This eventually led to the creation of Historic Fort Snelling, a reconstructed historic site operated by the Minnesota Historical Society, which is also a story of how we look at and try to understand the past. The paintings that start each chapter provide a visual record of the changing artistic perspectives on the fort.

This work is also intended to aid in our understanding of the past. Readers may appreciate not only the strictly military aspects of the fort's story but also how the army and its outpost reflected the values and institutions of the nation that created them. Fort Snelling should be seen ultimately not as a fixed memorial to a real or imagined past but as a channel for exploring the complexities and contradictions of that past. It is a story older and richer than we might expect.

Looking up the Mississippi River from the confluence,
1840s. Seth Eastman, the artist, served at Fort Snelling as
a lieutenant (1830–33) and as its commander (1840–49).
His watercolors provide a rich, if somewhat romanticized,
record of the area and the Dakota as he saw them. *MNHS*.

1. A Land and Its People

THE CONFLUENCE OF THE MISSISSIPPI AND Minnesota Rivers—Bdote—is the heart of the Dakota homeland. In 1805, when Lieutenant Zebulon Pike and his troops ascended the Mississippi, they were not entering an unpopulated wilderness. Fort Snelling was built on a limestone bluff at the center of a powerful, thriving nation.

The people living along the Minnesota and Mississippi Rivers had lived in the area for generations, sustaining themselves from the abundant resources in Mni Sota Makoce, the land where the water reflects the sky. Marriages and alliances created a web of relationships among the people. Trade networks extended across the continent. They knew the land thoroughly, as a relative. Dakota families spent their summers in villages placed along the Minnesota and Mississippi Rivers, growing corn and squash and gathering other foods and medicines. In the fall, they harvested wild rice on the region's abundant waterways and hunted deer; in winter, they lived in sheltered areas near their summer villages and fished with spears through holes in the ice. Spring brought women to the sugar bush to harvest maple sap and cook it into sugar, while men went to hunt muskrats. They used wood from the abundant forests for summer lodges, canoes, tools, and fuel. They visited each other for feasts, epic games of lacrosse, and ceremonies.[1]

The essence of Dakota life is to obey kinship rules, to be a good relative. Villages were made up of tiospaye, extended families. Men and women carried out separate and respected roles as partners in caring for the needs of the tiospaye. Working for your family, your village, and your oyate—your nation or people—brought health, respect, and prosperity for all. Relatives were expected to help provide for one another, sharing food and protecting other family members. To hoard food or accumulate goods for their own sake was not a sign of power but a mark of shame and boorish behavior. In fact, status was gained by giving away possessions through gift-giving and feasting. The Dakota saying "Mitakuye owasin" (we are all related) means exactly that: humans, animals, plants, and rocks exist in a network of relationships.[2]

Marriage and adoption required ceremonies that included gift-giving as an important part of the process, demonstrating the new member's commitment to his or her new relatives. Marriage, adoption, and gifts were important to intertribal

In his 1805 expedition, Lieutenant Zebulon Pike surveyed the limestone and sandstone bluffs along his route for potential locations for military posts. The story of these bluffs is a crucial part of the story of Fort Snelling, for the landscape determined where the fort would stand and how the area around it would develop.[i]

Some 500 million years ago, much of what is now the central part of North America was covered by a warm, shallow sea, accumulating layers of limestone, sandstone, and shale on its floor. Eventually, the central part of North America began to rise, and the sea disappeared, leaving a vast region of sedimentary rock. Because it lay in the center of the continent, away from areas of mountain formation,

St. Anthony Falls, about 1850.
Watercolor by Adolph Hoeffler, MNHS.

there was very little folding or warping, leaving a layered structure with alternating strata of sandstone, limestone, and shale. For a time, this was a landscape of rolling hills cut by gentle streams and a few rivers, including the predecessor of the Mississippi. It would have resembled southeastern Minnesota of today.[ii]

Then came the ice. Beginning about two million years ago, a series of vast, continental glaciers formed, advancing and retreating as many as eighteen times. Over thousands of years, the rock debris they carried filled existing river valleys, while seasonal meltwater flowing from the glaciers carried sediments and cut streams in new directions. The most recent episode, referred to as the Wisconsin glaciation, greatly impacted the current landscape of Minnesota. About 100,000 years ago, this ice sheet filled the former course of the Mississippi with glacial debris. As the glacier retreated, around 15,000 years ago, it left ridges of sand, gravel, and rocks, great blocks of ice that melted to form lakes, and meandering streams of meltwater.

The massive quantity of water flowing from the shrinking glacier fed large lakes that formed where ridges or moraines blocked this outflow. The largest of these covered most of what is now eastern North Dakota, Manitoba, Saskatchewan, and the western third of Minnesota—a freshwater sea flooding almost 123,500 square miles. Called Glacial Lake Agassiz by geologists, it broke through the moraine at its southern limit about 12,700 years ago near present-day Lake Traverse. The resulting river flooded through a landscape already carved by retreating ice, joining the Mississippi near the current location of Fort Snelling. Known to geologists as Glacial River Warren, this was a mighty waterway up to five miles wide and at times hundreds of feet deep. It quickly eroded the overlying glacial till to find its bottom on a hard layer of ancient limestone. About eight miles below the junction with the Mississippi, just east of modern St. Paul, the combined rivers ran off the limestone and over the filled-up course of an older river valley, in a short time scouring away the glacial deposits. Here a massive waterfall formed.[iii]

Plunging nearly two hundred feet, the falls became a mile-wide torrent with a deeper drop and a flow rate much larger than modern Niagara Falls. The falling water ate away at the softer sandstone that underlay the limestone, causing repeated rock falls and quickly eroding upstream. After four thousand years, the falls split at the confluence with the Minnesota. One branch eroded up the Minnesota River Valley to the southwest and soon devolved into rapids. The other continued working northward along the Mississippi. Its smaller flow of water over hard limestone let these falls survive to be known as Owamniyomni by the Dakota—and renamed the Falls of St. Anthony in 1680 by a passing priest.[iv]

While the falls were receding, the vast Lake Agassiz was also changing, finding new outlets to the north while dramatically shrinking in size. The only watery remnants of this great inland sea are lakes Winnipeg and Winnipegosis in Canada and Upper and Lower Red Lakes in Minnesota; the broad, flat Red River Valley is its former lake bed. The Minnesota River occupies the wide, deep valley carved by the River Warren, meandering across its spacious course like the solitary resident of an opulent mansion. Even the wider, swifter Mississippi does not quite fill the imprint left by its predecessor. Most evident today are the steep bluffs along both rivers, a landscape that has awed its human inhabitants for thousands of years.[v]

The River Warren Falls rampaging through the area near today's Fort Snelling must have been an overwhelming sight. Its rumbling and the smokelike mist rising from the base of the torrent would have been heard and seen from far away. This was a place of mystery and power, where the forces driving the universe revealed themselves, and here may mark the beginning of the site's sacred nature, for there were people here to witness it. ✍

The summer village of Medicine Bottle (Wakaŋ Ożaŋżaŋ), 1840s, located downstream from St. Paul at what is now Pine Bend. These Dakota seasonal lodges were constructed from local wood using elm bark and saplings. *Watercolor by Seth Eastman, MNHS.*

relations as well, sealing agreements with other tribes. Among their many other roles, women often played a central part in diplomacy, first as brides, then as wives and mediators between old relatives and new. Women who brought wisdom and good counsel to this role were honored in their communities. When contact and exchange began with Europeans, the Dakota extended this practice of intermarriage to this foreign people.[3]

The Dakota are part of the Oceti Śakowiŋ, the Seven Council Fires, known then to their Ojibwe neighbors as the Nadouessioux (like a snake) and to both Europeans and Americans as the Sioux. Among their origin stories is one that names Bdote as the place of creation: the people came to the earth from the seven stars in the constellation Orion. Four of the seven bands— Mdewakanton (Bdewakaŋtuŋwaŋ), Sisseton (Sisituŋwaŋ), Wahpekute (Waȟpekute), and Wahpeton (Waȟpetuŋwaŋ)—make up the eastern or Santee Dakota. The Yankton (Ihaŋktuŋwaŋ) and Yanktonai (Ihaŋktuŋwaŋna) are the western Dakota, and the Teton (Tituŋwaŋ), who live farther west, are commonly known as the Lakota.[4]

The Oceti Śakowiŋ have no migration story; their oral histories hold that Mni Sota Makoce is their homeland, their place of origin. Archaeologists also recognize that the people of the Oyate are descended from the earliest human inhabitants of the region. The scientists, however, place the arrival of those humans at about 12,000 years ago, at the end of the last ice age; they trace these migrant origins to the people who crossed Beringia, the land bridge that once connected Asia and North America, perhaps eight thousand years earlier.[5]

The Dakota eventually spread over a wide region from what is now western Wisconsin to the Dakotas and northern Iowa throughout Minnesota to western Ontario and eastern Manitoba.

The Yankton, Yanktonai, and Teton moved even farther west, drawn by the bison on the open plains in what is now South Dakota. Living in a region abundant in resources, the Oceti Śakowiŋ became a widespread and numerous people, sometimes conflicting with neighbors over those resources.[6]

The whole region around the confluence held—and holds—areas of spiritual significance for the entire Oyate but especially for the Dakota who lived there. The remnant of the River Warren Falls, the great waterfall on the Mississippi upstream from Bdote, was known as Owamniyomni and revered as the home of a powerful spirit. Early European accounts confirm their importance. In 1680, the Jesuit priest Louis Hennepin, who traveled through the area, renamed Owamniyomni the Falls of St. Anthony. He described a Dakota man offering a buffalo robe to the spirit of the falls, and a similar account was given by the Englishman Jonathan Carver a century later. Other places of importance included nearby Mni Sni (Coldwater Spring), Oheyawahi (The Hill Much Visited, or Pilot Knob), and Wakaŋ Tipi (Carver's Cave), the cave under the bluff where the city of St. Paul would grow.[7]

Dakota villages in the area included Little Crow's village, Kaposia (Kap'oźa), at the bend of the Mississippi in what is now St. Paul; Ohaŋska (Black Dog Village), on the south side of the St. Peter's River, as the Minnesota was then known, about six miles upstream from the confluence; Titaŋka Taŋnina (Penichon Village), Tiŋta Otuŋwe (Shakopee Village), and Maya Skadaŋ (White Bluff Village), even further up the St. Peter's; Maya Kiçaksa (Sleepy Eye Village), at the confluence of the St. Peter's and Cottonwood Rivers; He Mni Caŋ (Red Wing Village), on the Mississippi near its confluence with the Cannon River; and Kiyuksa (Wabasha Village), further down the Mississippi, at what is now Winona.[8]

By the late 1700s, the Dakota knew of the Americans, but they had far more experience with the French and British. And that experience had set up the expectations of both peoples who met at the confluence in 1805.

⁓

The Dakota had been trading with the French since the late 1600s, first through Ojibwe middlemen and later directly, through a number of posts the French established on the Upper Mississippi.

A misty morning near the confluence, 2016. *Photo by Matt Schmitt.*

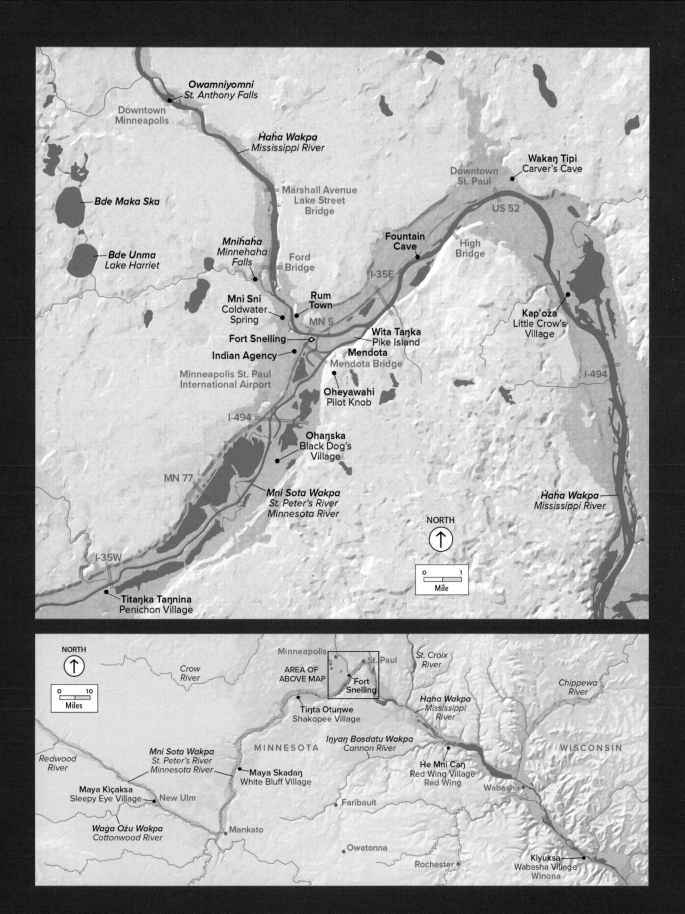

The cliffs near the site of Kaposia in 1805, shown in 2016. The area was known as Imniza Ska, white cliffs; the cliffs are at the base of what is now called Dayton's Bluff. *Photo by Peter DeCarlo.*

By exchanging furs for European trade goods, the Dakota and other Indigenous people gained access to materials they did not have the technology to produce themselves. These were primarily metal tools like knives, hatchets, and traps; domestic items like iron or tin cooking pots; textiles, particularly woolen blankets; and manufactured goods like glass beads and ribbons. While the traders may have considered these last items to be trinkets, for Indigenous people they were prized materials that were widely incorporated in regalia, sheaths, containers, and other equipment. Guns were also valued items, though in the seventeenth and eighteenth centuries firearms could be cumbersome and unreliable. Generally useless in wet weather, they were inaccurate at long range and required still more trade goods—gunpowder and lead—to keep them going. Still, a bullet from a trade musket could bring down a large animal or enemy warrior with a single well-placed shot, and their smoke and noise were impressive.[9]

The Dakota, like other tribes, desired these products, but their role in this international trade did not leave them dependent on their French partners. They still had the means to support themselves without European technology, and occasional interruptions in the trade, usually caused by European conflicts, did not create a disastrous situation for them. As long as they remained independent and the demand for furs continued, the Indigenous nations held an equal if not advantageous position in their relations with Europeans.[10]

For their part, the French found business success in the fur trade. From the onset, European colonization in the Americas was driven by a desire for wealth—Columbus, after all, was looking for a direct route to the rich trading regions of eastern Asia, not a continent previously unknown to Europeans. The French had relatively few riches to exploit in the lands they claimed around the mouth of the St. Lawrence River. Although they could practice some European-style agriculture there, nothing they grew compared to the cash crops—indigo, rice, and tobacco—developed in the English colonies. Furthermore, French colonial policy enforced a more restrictive emigration system, requiring loyalty to monarch and church, so the population of French colonists was smaller and less productive than New England's.[11]

The fur trade offered them a valuable opportunity. Europe's rising middle classes were eager customers for prestigious furs, most of which had come from rapidly depleted sources in Scandinavia and Russia. When the French opened up the St. Lawrence region, they discovered a vast source of furs in the northern forests and a Native population willing, even eager, to supply their merchants with this prized commodity in exchange for metal tools, cloth, and weapons easily manufactured in France. While the coats of ermine, mink, and otter were always welcome in the trade, it was the beaver that supplied the real prize. Its fur was valuable as a pelt, but when converted into felt, it could be fashioned into attractive hats that were both durable and waterproof.[12]

Through the 1600s, French merchants and their suppliers pushed ever farther into the interior of North America, seeking to claim lands for their esteemed monarch, but always in search of partners for their fur trade. Because there were not enough French soldiers or French colonists in North America to extend their power deep into the interior, the French had no means of coercing the tribes or enforcing their edicts. Indigenous men hunted and trapped; Indigenous women processed the hides. Without the Indigenous people, there would be no trade, and without Indigenous allies, the French could not hold on to New France itself. Only by withholding trade could the French exert any influence, and doing so could prove as painful to the royal treasury as to the peoples of the Great Lakes. Because continued trade depended on continued good relations, the colonial French paid careful attention to Native preferences, customs, and mores.[13]

The French had learned of the Dakota in the mid-1600s through intermediary Indigenous nations—the Huron, the Ottawa, and the Ojibwe. Under pressure from the warring and well-armed Iroquois, the Ojibwe had moved from the lower Great Lakes area to Sault Ste. Marie, where Lake Superior empties into Lake Michigan. For a time, this was an advantageous position, allowing the Ojibwe to act as middlemen between the French and western tribes, including the Dakota. In 1678, with the marriages of women who had accompanied an Ojibwe delegation to the Dakota village at Fond du Lac (near present-day Duluth) to young Dakota men, they established kinship bonds that reinforced the trade ties.[14]

A French emissary who also accompanied that Ojibwe delegation, Daniel Greysolon, Sieur Dulhut, traveled farther into Ojibwe country and spent the winter of 1679–80 at another large Dakota village at Mille Lacs (Bde Wakaŋ, or Spiritual/Sacred Lake). There he learned of the "captivity" of Father Hennepin, who had been traveling with a small party sent to explore the Mississippi by the expedition of René-Robert Cavelier, Sieur de La Salle, in February 1680. A Dakota war party found them two months later and took them home, perhaps recognizing their peril. Dulhut's "rescue" of Hennepin, who published an exaggerated account of the events, would influence European and later American perceptions of the Dakota and their homeland. As historian Mary Wingerd explains, "The complexities of cultural interaction were lost in a national myth of 'taming' the frontier that reduced Native Americans to alien savages who stood in the way of civilization.

Dakota woman preparing a hide, 1840s. *Sketch by Seth Eastman, MNHS.*

This version of the national saga so informed the consciousness of American culture that no room was left for alternative interpretations, even when, as in the case of Father Hennepin, the historical records called events into question."[15]

Yet the trade made important changes in both cultures in North America. If the French merchants wanted furs, they had to accommodate their customers' rules. This meant conforming to Dakota social norms, as they had to those of the Ojibwe. Frenchmen wishing to trade directly with the Dakota generally had to establish kinship ties through marriage à la façon du pays (after the custom of the country, using local rather than Christian ceremonies), providing gifts to their new relations; these relatives, in turn, helped their new kinsman acquire furs. A coureur des bois (runner of the woods) worked at the cultural intersection of the fur trade. Many of them were lower-class immigrants with little social standing in French society who embraced the Indigenous cultures that accepted them as equals, adopting them as members of their tribe and family. As such, they often acquired important survival skills and knowledge, including familiarity with Native languages. Thus, they became integral to the trade. The children of their marriages were still more essential, as they were fully bilingual with close ties to both cultures. Some of these individuals would become important leaders or advisors within their bands, guiding them in their relations with Europeans.[16]

This pattern of cultural accommodation established by French traders in the later seventeenth century persisted in the Upper Mississippi region until the early nineteenth century, surviving political upheavals, wars, and outside efforts to monopolize the trade. The region acquired a class of independent people of mixed ancestry who became critical for the fur trade and Indian-white relations in general. While European trade influenced material life for Indians, it was Indian social

customs that influenced the lives of Europeans and Americans on the Upper Great Lakes and in the Mississippi Valley for 150 years. This was a direct legacy of the French fur trade, as was the host of place-names that still appear in the landscape from Lac qui Parle to Prairie du Chien.[17]

A less happy result of French traders was the long, intermittent conflict between the Ojibwe and the Dakota. Other factors, like the continued migration of eastern tribes westward under pressure from British colonists on the East Coast, may have eventually led to this conflict. But it was actions of the French explorer and trader Pierre Gaultier de Varennes, Sieur de la Verendrye, that set it off. In 1731, as he sought to push the boundaries of New France westward and locate the ever-elusive passage to the Pacific, Verendrye generously distributed guns and metal weapons to the Cree and Assiniboine tribes and established a permanent trading post, Fort Charlotte, on Lake of the Woods. Knowing that the Cree were bitter enemies of the Dakota, he encouraged the authorities in Montreal to establish trading posts on Lake Pepin, to compensate the latter. Thus, the French found themselves arming both sides in an ongoing conflict between Indigenous nations.[18]

Matters escalated in 1736 when the Dakota took revenge for these acts, which they saw as a betrayal of the kinship relations they had established with the French. In June of that year, a largely Dakota war party attacked and killed a group of twenty-four Frenchmen, including the son of Verendrye, mutilating the bodies. In response, the colonial government went to war with the Dakota, furnishing arms to the Cree, Assiniboine, and Ojibwe. With a growing population and strong economic ties to the French, the Ojibwe were anxious to expand their territory, while the Dakota, less dependent on the French, saw no need to tolerate Ojibwe encroachment on their hunting grounds. So this persistent conflict, which would drag on for over a century, had its

Buffalo grazing near the mouth of the Minnesota River, 1840s. *Watercolor by Seth Eastman, MNHS.*

origins in the international fur trade and French colonial policy.[19]

About this time, the Dakota abandoned the region around the headwaters of the Mississippi and Mille Lacs, areas they had occupied for centuries. While fighting with the Ojibwe and their other enemies probably influenced the move, the Dakota were centering more of their activities to the south and west. Some bands already lived there, hunting buffalo on the prairie and trading at French posts on the Mississippi. Dakota bands lived in different landscapes, but they were still connected by the great river routes of the Mississippi and Minnesota. The confluence of these rivers was literally and spiritually the center of their world.[20]

Though the Upper Mississippi and its peoples were distant from Europe's homelands, the political struggles of the Great Powers and their colonies nonetheless influenced events in the region. The long struggle between England and France reached a turning point in the Seven Years War, known to England's American colonists as the French War and later as the French and Indian War. This was truly a worldwide conflict, which began with clashes between French and English colonists in the Ohio River Valley of North America in 1754–55. In the end, the English and their allies, the Iroquois Confederacy, were victorious, capturing Montreal and Quebec. The Treaty of Paris in 1763 officially ended the war and established Britain as a world power with an empire stretching from the Spice Islands of Asia to central North America. The French relinquished to Britain all their claims to North America east of the Mississippi, leaving the fur trade of the Great Lakes to British traders and merchants.[21]

The immediate effect of the war was to disrupt the trade in the Upper Mississippi region, which ironically brought a measure of peace among the tribes. Without the pressure to provide furs and with the interruption in the supplies of guns and ammunition, fighting between the Ojibwe and the Dakota declined. Sporadic fighting between individuals or families occurred from time to time, but so did intertribal marriages and alliances. Such important leaders as Flat Mouth (Eshkibagikoonzhe) among the Ojibwe and Wabasha (Wapaháṡa, The Leaf) of the Dakota were children of such marriages.[22]

Following its success, the British government tried to extend more direct control over the entire region west of the Allegheny Mountains, treating it as any other conquered territory, but the Indigenous nations there resisted. The resulting Pontiac's War nearly drove the British from the Ohio Valley and western Great Lakes. The government changed tactics, opening the region to any licensed trader. Though many entered the region from both the British Isles and American colonies, they relied on the existing French-speaking traders of mixed ancestry and their Native relatives to carry on the trade. In fact, with a few exceptions, both the English and their American colonists had a decidedly aloof attitude toward their partners, regarding both Indigenous people and those of mixed ancestry as social and racial inferiors. Yet these children of the fur trade, with cultural roots in both the Indigenous and European worlds, would remain entrenched in the trade and the region as indispensable intermediaries well into the nineteenth century.[23]

Through the 1760s and 1770s, the trade was fluid, with many individuals competing for access to furs. The tribes occupied a position of extraordinary leverage, expecting substantial gifts from traders in return for furs. This situation was difficult for traders, particularly as the American Revolution created a shortage of goods for independent traders to exchange. Montreal merchants, now dominated by Scottish immigrants to British Canada, took advantage of the situation to organize the trade. The North West Company (NWC), founded in 1779, created a vertical monopoly, locking up the supply of European goods and transportation of those items into the Great Lakes. The company forced independent traders to join as "wintering partners" or be ruined. They thus succeeded in controlling the trade into western Lake Superior and the Rainy River country, with Grand Portage serving as their primary base of operations. From this dominant economic position, the NWC compelled the Ojibwe to accept a far less favorable situation, where the importance of kinship bonds was diminished.[24]

The Dakota and their southern neighbors, the Sauk and Meskwaki (Sac and Fox), remained independent, but a few traders were able to establish prominent places among them. Notable examples were Robert Dickson and Joseph Renville. Dickson, a Scot, was among the few British traders to recognize the importance of making direct connections. He married Totowiŋ, a daughter of Sisseton chief Red Thunder. Renville, a French Canadian trader, married into Red Wing's band of Mdewakanton Dakota. Relationships like these firmly established the British among the Dakota, who continued to value European goods.[25]

While the Dakota were not as affected as the Ojibwe by the monopolization of the fur trade, they did not escape the impact of the smallpox epidemic that ravaged most of North America during the American Revolution. The overall population of the Oceti Ṡakowiŋ in 1780 is estimated to have been about 25,000; by 1786, the four eastern bands that made up the Dakota were estimated at 4,200, suggesting a serious toll. Yet the Dakota remained an important ally for whoever

This silver trade cross, five inches tall and two and three-quarters inches wide, was made by silversmith Robert Cruickshank of Montreal between 1767 and 1809. It was found at Sandy Lake, Minnesota, near where a North West Company trading post was established in 1794. *MNHS*.

would befriend them, and in the immediate aftermath of American independence, British traders, most notably Dickson, were determined to stay.[26]

Another aspect of British trade, however, was more complex and troublesome. While hunters needed gunpowder and lead and repairs to guns and traps, a family could carry only so many kettles. Alcohol, usually in the form of rum or whiskey, became increasingly important as "the pernicious glue," in the words of historian Mary Wingerd, that held the trade together. It was the one item always in demand from Native customers. The tendency of alcohol to soothe aches and pains, as well as producing feelings of euphoria that created a connection to the spirit world, contributed to its popularity. Whites also consumed alcohol in large quantities, for many of the same reasons. In a world of hard labor in often harsh conditions, rum and whiskey were looked on by traders and soldiers as a necessity of daily life.[27]

For Indigenous people, the consumption of alcohol was less about a notion of necessity and more about how it was made available to them. Since they moved from season to season within their territories, people transported only what they needed to use. A barrel of rum given to a leader was likely to be immediately distributed to his followers, both as a sign of his readiness to share and as a practical necessity: there were no easy means to transport it. As the addictive quality of alcohol became more apparent among the tribes, the cutthroat competition of the fur trade drove traders to use greater and greater quantities of liquor to entice customers to sell their furs, often luring them away from other traders. A vicious pattern was created, centered on the toxic appeal of alcohol.

British success in the Seven Years War had given the empire nominal claim to a large part of North America, but the victory also led to conflict with the American colonies. New British taxes, some imposed to pay for the war, and policies restricting the westward expansion of the British colonies led to conflict between the British government and its colonial assemblies. Within twelve years, a full-scale rebellion was underway, leading eventually to American independence and the establishment of the United States. Britain maintained its influence on the fur trade in the Great Lakes and Upper Mississippi regions into the early nineteenth century, but a rapidly expanding American presence eventually brought dramatic changes to the region.[28]

Pilot Knob rises beyond Pike Island at the confluence, 1840s. *Watercolor by Seth Eastman, MNHS.*

2. The Arrival of the Americans

AT FIRST, THE CLAIM OF THE UNITED STATES to the West had very little effect. Native nations, residents for millennia, continued to live as they had. The British possessed a well-organized trading system and great influence, and they continued to trade. Yet American settlers were already trickling into the Ohio River Valley, and two political agreements between Europeans and Americans would shift the diplomatic realities enough to turn the trickle of immigrants into a stream. The first was the Treaty of Amity, Commerce, and Navigation of 1794, known as Jay's Treaty, which settled to some degree the nebulous boundary between British Canada and the United States. The other was the Louisiana Purchase in 1803, which gave Americans access to a vast new territory and important trade centers in St. Louis and, above all, New Orleans.

The United States had trouble establishing its influence over so large a region. The American states were badly in debt, poorly organized, and far away, with no military or political means of influencing events on the Upper Mississippi. Following the Revolution, the American army had been all but disbanded. Many Americans distrusted the concept of a permanent standing army as a burdensome expense and possible instrument of tyranny. They preferred the idea of a "well regulated militia" as expressed in the Second Amendment to the Constitution: a force of citizen soldiers who could be relied on to come to the defense of the country in the event of foreign invasion.

It was a lofty, even romantic theory of national defense that quickly proved inadequate, particularly on the western edge of its territory, the borderland regarded as the frontier. A small, poorly organized force could not stop white Americans from illegally invading Indigenous lands or defend such interlopers from retaliation. Nor could it discourage British traders from continuing to operate in regions claimed by the United States. In fact, British troops still occupied posts at Detroit and Mackinac Island on the Great Lakes and actively encouraged Indian resistance to the Americans.[1]

The Americans tried to counter this influence by establishing forts in the Ohio Valley, along the Ohio and Wabash Rivers. They were opposed by a coalition of Algonquin-speaking Native American tribes, primarily the Shawnee (Shaawanwaki), Delaware (Lenape), Potawatomi (Bodéwadmi), and Miami (Myaamia), led by the Miami chief Little Turtle (Mihšihkinaahkwa). In 1791, a force of American troops, many of them militia, under the command of General Arthur St. Clair, was sent into the area of the Wabash, where they were surprised and decisively defeated by the nearly one thousand warriors of this coalition. In response, the Washington administration under the newly constituted federal government established an "American Legion": a small but well-disciplined force under General Anthony Wayne, which—along with a force of Kentucky militia—defeated Little Turtle's forces at the Battle of Fallen Timbers in August 1794. Wayne built a line of forts that succeeded in establishing power over the Ohio River region following his victory. Jay's Treaty, signed in 1794, recognized American control and left the Native peoples of the region no recourse but to make treaties with the United States.

Over the next decade, there would be various proposals to expand the US Army to control the frontier. President George Washington's secretary of war, Henry Knox, advocated a policy of "peace through justice" with the Indian tribes. He noted at the time of his retirement in 1794 that American policy in the past had been more destructive than the Spanish conquest of Mexico and Peru and a blight on the nation's reputation. Much of the conflict with Indian nations had arisen from unrestrained encroachment on Native lands by whites, often in violation of previous agreements. Knox advocated the creation of a line of frontier posts manned by regular army troops to police relations between Native people and whites. He further sought to regulate trade by establishing government agents "to reside in the principal Indian towns," to create what he called "an honorable tranquility of the frontiers." Knox's well-meaning proposal was largely ignored by a Congress unwilling to fund it or to risk potential political backlash from expansion-minded constituents. In addition,

some of its members were all too eager to profit from land speculation. Similar proposals would continue to come forth from the War Department, and all would meet a similar fate.[2]

By the time of the Louisiana Purchase in 1803, the administration of President Thomas Jefferson had reduced the army to little more than three thousand men scattered in a handful of posts from Fort Mackinac and Detroit on the Great Lakes to Fort Massac on the Ohio River and Fort Adams on the Lower Mississippi. The acquisition of the vast new territory of Louisiana created as many challenges as opportunities for the American government. The United States had struggled to establish its authority over the Ohio Valley; how would it gain control over so immense a region with a modest military operating at a great distance from its centers of support? In these respects, the government of Thomas Jefferson was little different from those of France and Britain, too distant and too weak to impose direct control. Their influence would depend on the goodwill of Native peoples of the region. Yet the tribes of the Northwest had little incentive to work with the Americans. The Dakota and their neighbors had long-standing relationships with the British. Their chiefs still wore the medals and carried the flags of King George given them by British traders. What they knew of the American "long knives" was not flattering. They viewed the Americans as invaders who would take their lands as they had done to the tribes in the Ohio Valley.[3]

Initially, Jefferson and his immediate successors adopted a policy of conciliation toward Indian people. They felt that the economic ties of trade would be more likely to win them over than a show of military force and would be less of a burden on the national treasury. And trade offered a means of taking the land: "To promote this disposition to exchange lands which they [Native Americans] have to spare and we want," President Jefferson wrote in 1803 to William Henry

Harrison, governor of Ohio Territory, "We shall push our trading houses, and be glad to see the good and influential individuals among them run in debt, because we observe that when these debts get beyond what the individuals can pay, they become willing to lop them off by a cession of lands." Jefferson instructed the first governor of Louisiana Territory, General James Wilkinson, to follow a policy of encouraging peace with the tribes while observing "strict economy." Wilkinson was a controversial figure. In the freewheeling environment of the Louisiana frontier, he often acted more in his own interests than in those of the United States, involving himself in a number of shady incidents. Most notable was his participation in the Aaron Burr conspiracy—through which he was also acting as a paid secret agent of the Spanish government. Nonetheless, he clearly grasped the challenges facing the Americans in the Upper Mississippi region; British traders still dominated the area, and their influence with the Indians remained uncontested.[4]

In August 1805, Wilkinson took several actions intended to alter this situation. He issued a proclamation forbidding "foreign" (i.e., British) trade on the Upper Mississippi and Missouri Rivers. He also sought to locate forts in the region to enforce this policy and steer trade away from the Great Lakes and toward New Orleans. With these objects in mind, he ordered a reconnaissance of the Upper Mississippi to locate the river's source, determine the extent of British trade, and scout locations for military posts, among other things.[5]

To lead this expedition, Wilkinson selected an ambitious young officer named Zebulon Pike. Born in 1779, the son of a Continental Army officer, Pike represented a newer generation of American army officers. Self-educated, influenced by both English and American ideals of manhood and military honor, he was personally ambitious

Zebulon Pike, about 1810. *Engraving from* Analectic Magazine *4, no. 353 (November 1814), MNHS.*

yet zealously devoted to the welfare of his newly founded nation.[6]

Early in his career, Pike had become something of a protégé of General Wilkinson, whom he regarded as a father figure. Commissioned as a second lieutenant in March 1799, Pike earned early assignments that placed him in a wide range of duties, often in charge of small detachments delivering supplies to frontier posts. These were unpleasant, thankless tasks, exposing Pike to the vexations of river travel, including insects, wind, rain, and unpredictable water levels. He also occasionally found himself acting as a courier for Wilkinson to high-ranking officials in Washington. Pike's background, connections, and enthusiasm made him an ideal choice to command this expedition.[7]

While at Fort Kaskaskia in 1803, Pike had met

The Minnesota River flows into the Mississippi River, looking downstream from Wita Taŋka or Pike Island, 2016. *Photo by Matt Schmitt.*

Meriwether Lewis and William Clark, whose Corps of Discovery was beginning its epic journey of exploration to the Pacific. A number of men were recruited from the post to join the expedition, and Pike began a correspondence with Lewis, who suggested that Pike might lead his own exploration into Louisiana Territory. A year later, in the spring of 1805, General Wilkinson ordered Pike to command just such a reconnaissance on the Upper Mississippi.[8]

There was something quixotic about Pike's mission. Wilkinson's orders were extensive and complex, more suited to a larger, better-equipped party than the young lieutenant commanded. He was to locate the source of the Mississippi River, an important point for determining the border between the United States and British Canada;

reassure the Native peoples, negotiating treaties with them to locate forts on Indian territory; and all the while record the natural resources of the country along the route. Just locating the source of the Mississippi would have been a major accomplishment in itself, but the ever-zealous Lieutenant Pike added his own determination to show the flag and press American territorial claims against the well-established British traders.

The expedition was hampered by the fact that they embarked very late in the season, leaving St. Louis on August 9, 1805, in a large, heavily laden keelboat. Low water on the river further delayed their progress, and they did not reach the French settlement at Prairie du Chien until September 4. Here the party exchanged its keelboat for smaller craft to facilitate their journey upriver. Pike's party

also gained some important new members: several traders of both European and Indigenous descent, then known as mixed bloods. These men included Joseph Renville, who acted as Pike's translator.

After surveying the area for a location for a post, Pike continued upriver. He soon held his first meeting with a Dakota leader, visiting Wabasha's village on the Mississippi. Here he was feasted and, in an elaborate ceremony lasting several hours, made a relative to this prominent Dakota chief. He assured the Dakota of the good intentions of the American government and its desire to supplant the British and offer the Dakota better prices for their furs. He made stops at other Dakota villages on his way upriver, eventually arriving at the mouth of the St. Peter's River on September 23.

Local Dakota leaders were aware of the expedition and agreed to meet him on the island at the confluence of the rivers. Among them were Little Crow (the grandfather of the later leader by that name), Red Wing, Standing Moose, Shakopee, Walking Buffalo, and Broken Arm, along with a large number of warriors, reportedly as many as two hundred—an impressive assembly. Most of these leaders came from nearby Mdewakanton villages in the region surrounding Bdote, but Kaposia, Black Dog Village, Penichon Village, and Shakopee Village would eventually be the populations most directly impacted by the creation of the military post that became Fort Snelling.[9]

The result of the negotiations that followed was an agreement that Americans and Dakotas would come to understand in very different terms. Pike's

Little Crow, the leader of the Dakota village of Kaposia near Bdote in 1805. *Lithograph by Charles Bird King, 1836, MNHS.*

written agreement with the Dakota had them grant two large tracts of land: nine square miles at the mouth of the St. Croix River and another tract stretching nine miles on either side of the Mississippi River from its confluence with the St. Peter's to the Falls of St. Anthony. Pike later estimated the value of this land at $200,000. The United States would allow the Dakota to hunt and use the area as they were accustomed to, but would retain unrestricted use of and access to the area. The Dakota, who considered the natural world to be a relative rather than a commodity that might be sold, probably understood that they were allowing the Americans to use the land visible from the high ground along the river, amounting to a mile or two on either side.[10]

Little Crow and He Sees Standing Up (Waŋyaga Inażiŋ, also known as Fils de Penichon), leaders of the two nearby villages, signed the document. In return, Pike gave modest presents to the chiefs and promised a larger gift to the Dakota people later; the amount to be paid was left blank.

More important, he promised that the American "father" would soon establish a permanent trading post in their midst. (Europeans and Americans consistently used the language of relationships, so crucial to Indigenous diplomacy, to create tools that matched their goals of patriarchal domination and dependency.) Though there had been intermittent trading outfits established in their territory, the Dakota had long sought a more permanent, direct connection, something akin to the emporium at Mackinac the Ojibwe benefited from.[11]

The agreement was fraught with difficulties, not the least being that Pike had no authority from the US government to make a treaty. Wilkinson had sent off this expedition purely on his own account, and although its goals generally fit within government policy, there was no assurance that anything Lieutenant Pike proposed would be approved. Indeed, the US Senate did not take up the treaty until 1808. When discussing the agreement, they valued the land at just $2,000. This was in accordance with a committee report that calculated the total amount of land "ceded" at the mouth of the St. Croix and between St. Anthony Falls and the mouth of the Minnesota at 103,680 acres, which they valued at two cents per acre per government policy. President Jefferson never officially "proclaimed" the treaty, the final step in a formal treaty process—though the army later acted in accordance with its provisions.[12]

On the whole, the expedition was at best a mixed success. Pike did not locate the headwaters of the Mississippi, and he was himself rescued during the winter of 1806-7 by some of the British traders he was sent to confront. For a man who had spent most of his career, indeed a good part of his life, on the boundaries of the United States, it is amazing how unprepared Pike was for exploring the country beyond Prairie du Chien. He was clearly overconfident and had no compunction about risking the lives of his command. Further, the lieutenant seems to have regarded the region as an empty wilderness. How surprising it must have been for him to find British traders and traders of mixed ancestry, with their Indian allies and relatives, moving through the country with ease and living there in relative comfort. It might have been a disturbing revelation to find the region in a sense "settled" and under British control. Despite the traders' hospitality, Pike arrogantly accused them of trespassing on US territory in violation of Wilkinson's proclamation and demanded they allow him to examine their accounts. He even ordered the handful of soldiers with him to shoot down the British flag flying above the post. The traders politely agreed to his demands and helped him on his way, but as soon as the American was gone, they continued their trade as they had for the previous half century.[13]

Pike returned to St. Louis the following spring and soon published a glowing report of his achievements. Then he was off on another strenuous, less successful adventure that would leave his name on a peak in Colorado. Yet the expedition did succeed in revealing the extent of British trade on the Upper Mississippi and establishing US relations with the Dakota. For those relations to grow, the government would have to follow through on the assurances Pike offered regarding payment and, above all, trade. But during the next eight years, the United States did little to improve its influence in the Upper Mississippi country. President Jefferson's continued reliance on commerce, rather than a military presence, to influence the region's tribes assumed that the Americans could control access to the trade routes of the area. Instead, the British remained firmly established, regularly intruding into the Lake Superior and Upper Mississippi region.[14]

When war with Britain eventually came in 1812, it proved a near disaster for American interests in the West and Northwest. Initially, a considerable number of Dakota, mostly from the Mdewakanton band, fought as allies of the British, helping them overrun many of the key American forts in the area. (Pike himself was killed at the Battle of York in 1813.) In fact, had the war ended in 1813, the entire region would probably have returned to British control. As the war dragged on, however, decisive American victories at the battle of the Thames in the North and Horseshoe Bend in the South dramatically changed the situation in the Mississippi Valley. The Dakota became increasingly dissatisfied with the British, and many withdrew from the fighting. The Treaty of Ghent, which officially ended the war in December 1814, left unchanged most of the issues that had caused the war, and in doing so completely ignored the claims of British traders and their Indian allies against the Americans. The British government, its resources strained by the Napoleonic Wars, essentially surrendered influence in the Northwest to the United States, and this time the Americans would not wait long to establish themselves there.[15]

Fort Snelling, 1833. *Sketch by Seth Eastman, MNHS; gift of Frederick F. Seely.*

3. Building a Fort

FOR THE UNITED STATES, THE YEARS IMMEDI-ately following the War of 1812 were a period of national unity and pride—an "Era of Good Feelings." James Monroe was elected president in 1816, practically without opposition, and party divisions were at a low ebb. The country had fought the world's greatest power, Great Britain, to a standstill and remained independent of European dominance. In fact, it had come to see itself as the predominant power in the Western Hemisphere. Whether this outpouring of nationalism was justified by America's military might or Europe's indifference is questionable, but the fact remained that the United States still faced challenges in extending its influence into Louisiana Territory, especially the Upper Missouri and Mississippi Rivers.

The army was still relatively small and scattered. Though it had regained its western outposts in the Treaty of Ghent, they were weakly held, and again British traders were operating in their old haunts on the Upper Mississippi, like Big Sandy Lake. If anything, it was the influx of American and European immigrants rapidly colonizing the Ohio River Valley that extended American influence and simultaneously compelled the government to act. The region's military commander, General Jacob Brown, wrote to Secretary of War John C. Calhoun, noting, "Since the period of assuming the command of my present Division the frontiers of it have undergone an important change. The enterprise of our citizens has thrown into the interior forts which were originally intended for the defense of their frontier, [and] by enlarging the field of immigration, have removed the causes for which they were first erected."[1]

In fact, Brown had perhaps anticipated his superior's wishes. Though later known as a champion of states' rights and an ardent defender of slavery, John C. Calhoun began his career as an enthusiastic nationalist. As a young congressman from South Carolina, he was a leader of the War Hawk faction, supporting war with Great Britain. President James Monroe appointed him secretary of war in October 1817, and though he was not a military man, he worked quickly and effectively to reorganize the service. Calhoun, with Monroe's support, proposed a very ambitious plan for the frontier, designed to end once and for all British influence and to secure the region and its trade for the United States. He wanted the army to establish posts as far as the Mandan villages and the mouth of the Yellowstone on the Missouri and at the junction with the St. Peter's on the Mississippi. In his instructions to Brown, Calhoun was particularly specific in his wishes regarding the Mississippi.

> The two principal positions on the Mississippi will be at the junction of the St. Peter's and Fort Armstrong. The latter position is an Island [now Rock Island, Illinois] in the river and is said to be by nature very strong. The former, from its remoteness from our settlements, its propinquity to Gov. Selkirk's establishment at Red River of Lake Winnipeg and from its neighborhood to the powerful nations of the Sioux, ought to be made very strong. The force sent in the first instance ought to be as imposing as it can be rendered. With this view, it is thought advisable, to occupy the post on the Mississippi with an entire regiment [up to a thousand troops], the whole of which, with the exception of what will be necessary to garrison Fort Armstrong, and Fort Crawford at Prairie du Chien, ought to be moved up to the St. Peter's.[2]

The entire enterprise of establishing these far-flung military posts would prove a test of the army's and the US government's capabilities. Indeed, to support the movement and supply of troops, the War Department paid contractors "upwards of $250,000," a sum so great that Calhoun feared that "Those who are opposed to the Army will unquestionably seize the present embarrassment to magnify all of its expenses in order to reduce the establishment." This sum included the purchase of a steamboat, a relatively new technology in 1819, as well as keelboats and supplies. The bulk of this expenditure, however, was for the secretary's much-desired but unsuccessful Yellowstone Expedition, intended to establish the fort at that river's

Fort Armstrong, now Rock Island, Illinois. *Watercolor by Seth Eastman, 1848, MNHS.*

junction with the Missouri. But the regiment assigned the task, the 1st Infantry, commanded by Colonel Henry Atkinson, was not able to move up the Missouri beyond what is now Council Bluffs, Iowa.[3]

For the establishment of the St. Peter's post, General Brown selected the 5th Infantry Regiment, then stationed at Detroit. It was provisionally commanded by Lieutenant Colonel Henry Leavenworth, a native of Connecticut who had a distinguished record as an officer in the War of 1812. He had been lieutenant colonel of the 5th Regiment since 1818 and in command since early 1819, when his superior, Colonel James Miller, was promoted to brevet major general in charge of the 2nd Military District. His posting at Detroit gave Colonel Leavenworth some familiarity with the region and its challenges. Accordingly, he and three hundred men of the 5th Regiment left Detroit on May 14, 1819. Traveling via Green Bay and the well-established trade route connecting the Fox River to the Wisconsin River, they arrived at Prairie du Chien and Fort Crawford on June 30. Here they were to rendezvous with a flotilla of transports bearing supplies and new recruits sent up the Mississippi from St. Louis. These had not arrived, and Colonel Leavenworth soon learned that delays by contractors and the ever-changing nature of the great river would impede the 5th Regiment's mission.

A few days later, on July 5, Thomas Forsyth, Indian agent to the Sauk and Meskwaki, arrived in Prairie du Chien from St. Louis to accompany

Lieutenant Colonel Henry Leavenworth, about 1820. *Anonymous engraving, MNHS.*

Leavenworth's command. His journals note that the St. Louis contractors had given up the idea of sending a steamboat with the 5th Infantry's supplies and recruits. Two sets of rapids on the Upper Mississippi—one at the mouth of the Des Moines River, near present-day Keokuk, Iowa, and the other at Rock Island—created problems for steamboat navigation. The contractors instead hired keelboats to do the hauling. Forsyth had hired his own boats and crew to carry himself and supplies, including "a certain quantity of goods, say $2,000 worth, to be delivered by me to the Sioux Indians residing on the Mississippi above Prairie du Chien, and those who reside on the lower part of the St. Peter's, in payment of lands ceded by the Sioux Indians to the late Gen. Pike." This amount matched the amount set by the Senate as payment for the land. The journey from St. Louis took nearly a month because of high water, strong currents, and contrary winds.[4]

Another month of waiting followed, and not until early August did the supply boats begin to arrive. They reported passing a boat or two with recruits headed north, but by August 7 these had not come in. Meanwhile, Colonel Leavenworth hired several boats at Prairie du Chien to take his troops and supplies upriver. Having experience in the Northwest, he knew there was little time to complete his mission before winter would make it impossible. Leaving a detachment of men at Fort Crawford, Leavenworth proceeded upriver, arriving at the mouth of the St. Peter's on August 8,

Landscape near Wabasha's village. *Watercolor by Seth Eastman, 1840, MNHS.*

1819. As Forsyth described it, the expedition included "fourteen bateau [small wooden boats, similar to canoes] and two large boats, loaded with provisions and ordnance, and stores of different kinds, as also my boat and a barge belonging to the colonel making seventeen boats; and in the whole, 98 soldiers and 20 boatmen."[5]

An important part of Forsyth's role, and the primary purpose of his presence with the expedition, was to smooth their way with the Dakota. Though the government had made peace with some bands following the War of 1812, the Americans knew that the Dakota were unhappy with the Americans' failure to follow up on the promises Pike had made in 1805. A few days after departing from Prairie du Chien, the troops reached the village of chief Wabasha, whom Forsyth in his account referred to by the translation of his name, "The Leaf." This was the same chief Wabasha that Pike had met with fourteen years earlier, and he was still one of the most influential leaders among the Dakota.

The agent expounded at length the purpose of the soldiers and their good intentions. He explained the advantages for Wabasha and his people in having a fort located so strategically near them. There would be opportunities for trade, and the blacksmith located there could repair "any little thing" for them. He further assured them that the soldiers would protect them from their enemies anytime they visited the post, and the Dakota in turn must promise not to attack any visiting Ojibwe. He further assured Wabasha "that if their Great Father, the President, meant them any harm, he would not send a man of my years having so many grey hairs in his head as I have, to do anything but what is good." He also promised that if the Dakota had any complaints against any of the soldiers at the fort, they had only to mention them to Colonel Leavenworth: "He will render you every justice in his power, and both him and myself will expect that if any

of your young men do what is not right, you as the head chief, will render justice equally in the same way when the colonel complains to you." Wabasha accepted Forsyth's small gift with thanks and later visited the agent on his boat, "and we conversed on many subjects."[6]

The expedition continued upstream, stopping at the villages of Red Wing and Little Crow along the way. Forsyth was particularly impressed by Little Crow, who had also been present at the council with Pike in 1805. "His independent manner I like," wrote the agent. "I made him a very handsome present, for which he was very thankful, and said it was more than he expected." This extra-generous gift may have been part of the long-delayed treaty payment.[7]

On August 23, the troops reached the area of Bdote, and on the following day, after much scouting for a good location, they chose a spot down in the river valley, "immediately on the mouth of the St. Peter's River, on its right bank." The soldiers were immediately put to work constructing winter quarters, which Colonel Leavenworth christened Cantonment New Hope. The name would soon prove ironic.[8]

Using troops for the construction of fortifications and roads had been War Department policy since 1815. The men received additional pay of fourteen or fifteen cents per day and an extra whiskey ration for their labor. Given the isolation of many of the new posts and the difficulties of supply, this seemed a logical arrangement, even a popular one. As General Jacob Brown pointed out, "It did not occur to me that it would be thought unreasonable to allow the soldiers so employed extra spirits and also extra pay—I have no doubt but that it will be found true economy to pay the soldiers for their labor agreeable to the former regulations but at a higher rate as labor of all kinds is much higher than at the time the regulations I refer to were established. I would say 14 cents per day for common laborer especially and 20 for mechanics—This would be an encouragement to the troops and greatly promote the views of Government."[9]

In the meantime, Agent Forsyth met with a number of Dakota leaders who had traveled down the St. Peter's to greet him and receive gifts similar to those he had been dispensing. He gave them all the same assurances he had given Wabasha, Red Wing, and Little Crow regarding the new fort, and after distributing the remainder of his goods, he returned to St. Louis. He was accompanied as far as Prairie du Chien by Colonel Leavenworth, who was looking out for his much-needed and long-delayed supplies and recruits.[10]

Work on the new encampment was put under the charge of Major Josiah Vose, who quickly pressed forward with the building of a substantial post. The plans apparently were Leavenworth's, though Vose modified them somewhat,

concentrating on completing the quarters. In his report to the colonel, Major Vose described the completion or near completion of some forty-five rooms, including quarters for officers (some with families), noncommissioned officers, and enlisted men, along with storerooms, kitchens, two sally ports (small exit gates or doors for troops making a sortie), and a guard room. Most of the living quarters even included sash windows with glass panes. He expected the quarters to be habitable by mid-November, while the construction of a proper powder magazine and blacksmith shop would have to wait.[11]

Vose credited much of the work to two fellow officers, both of whom would later figure in the building of the permanent fort: Lieutenant Robert A. McCabe and Major Joseph Plympton. "Lieutenant McCabe," he wrote, "has superintended the quarrying of stone and building of Chimneys. He

Henry Rowe Schoolcraft, traveling down the Mississippi with the Lewis Cass expedition in 1820, sketched Cantonment New Hope in his journal. He labeled the cornfield and garden, in foreground; the structures at center left include the barracks, the long building. Pike Island is at left. *Henry Rowe Schoolcraft Papers, MNHS.*

has also erected a lime kiln and made some lime." Plympton had the still-greater task of "the principal director of the buildings and charge of the several parties and is entitled to much credit for his indefatigable attention and perseverance in procuring timber, directing the workmen, etc. etc." The major also noted, "It has been found extremely difficult to obtain timber, and particularly such as was suitable for shingling." Scarcity of usable wood for building or for fuel proved a continuing challenge for many years.[12]

This was a prodigious accomplishment for the two hundred or so men, including 120 recruits who arrived in September, especially as many of them were suffering from "ague and fever." According to Major Vose, "The sick report since the arrival of the recruits has been large, averaging forty to sixty, four of the recruits have died." Several diseases, including influenza and food poisoning, may have been the culprits in this outbreak, but the most likely affliction was malaria. This would not be its last appearance.[13]

For all this expenditure of effort and treasure, the whole expedition started to seem like a stretch to Henry Leavenworth. His report to Major General Alexander Macomb, commander of the army's western division, reflected anguish and frustration. He stated frankly that the provisions so laboriously rowed, poled, and hoisted up the Mississippi that summer would not last through the following February. In fact, the government had paid for shipping the supplies only as far as Prairie du Chien and still owed $2,000 to contractors. He was also becoming acutely aware of the 5th Regiment's isolation. "I every day feel more and more sensibly the want of instruction and order from my government through my commanding officer. It will soon be necessary to commence obtaining materials for a fort, and I have not yet been told what kind of work to build or that it was left to my discretion." Further, the repeated transfer of men from the regiment and the nature of their mission was affecting morale. "Our officers are thus subjected to the expense and trouble of recruiting men of other corps and to the mortification of hearing their recruits complain of being deceived as to the Corps and nature of service for which they enlisted."[14]

If Leavenworth's message seems tentative and indecisive for someone entrusted with an independent command, it should be remembered that Henry Leavenworth's position was somewhat tenuous. Though he was the officer in charge, he was not officially the commander of the 5th Regiment. He had been the regiment's lieutenant colonel, technically second in command, since 1818, so he was actually in temporary command, which may account for his anxiety and request for instructions.[15]

Leavenworth received a direct reply from Secretary of War Calhoun, who praised the efforts of the 5th Regiment. "I am much gratified to find that your movement has been made without opposition or hostility on the part of the Indians, and the promptness and judgment with which it has been executed is highly satisfactory to the Department," the secretary wrote.

> The military movement such as has been made up the Mississippi under your command was ordered for the establishment of posts to effect two great objects, the enlargement and protection of our fur trade, and the permanent peace of our North Western frontier, by securing a decided control over the various tribes of Indians in that quarter. . . . To such of our citizens who conform to the laws and regulations in relation to Indian trade and intercourse you will extend kindness and protection. In relation to foreign traders, who by the act of congress, are entirely excluded, your conduct in the first instance must be governed by a sound discretion to be exercised in each case. No decisive step ought perhaps to be taken until your posts are fully established and you feel yourself secure against the effects of hostilities, at which time notice ought to be given that after a fixed period, you will rigidly

exclude all trade by foreigners and such as are not authorized by law. Of the two great objects in view, the permanent security of our frontier is considered by far of the greatest importance and will especially claim your attention.

He also informed the colonel that an Indian agent, Lawrence Taliaferro, would soon depart Washington for the St. Peter's to assist with these wide-ranging duties.[16]

Though the regiment completed the building of reasonable shelter, the 5th Infantry's first winter in Minnesota was a dismal experience. Low water on the Mississippi prevented the arrival of critical food supplies, and in addition to the diseases of autumn, the winter months saw many of the soldiers stricken with another severe illness, later described as a particularly virulent form of scurvy. There were accounts of men dying in their sleep during the night and even soldiers on guard duty found dead when their relief arrived a few hours later. Though official accounts generally blamed scurvy, that disease works slowly and seems unlikely to have caused such sudden deaths. Philander Prescott, who joined them the next summer to work with the sutler, noted that they were eating only bread and "rusty pork." Contractors transporting rations for the troops had removed the salt brine from the barrels of salt pork to lighten the load in their keelboat and so help it navigate the shallow areas on the river. Later they refilled the barrels with river water before arriving at Prairie du Chien. So the soldiers' salt pork ration was of dubious quality.[17]

The following spring, Colonel Leavenworth decided to move the new post to a better, healthier location high above the Mississippi at the spring the Dakota called Mni Sni, Coldwater Spring, about a mile and a half to the northwest. Like the rest of the Bdote area, the spring was and is a sacred site to the Dakota, a dwelling place of Uŋktehi, the water spirit. During the summer of 1820, the troops, still recovering from scurvy, made slow

headway building the new camp, known as Camp Coldwater. Living in temporary huts, they worked on building storehouses while beginning to grow crops "under orders of the War Department which is absolutely necessary to preserve them from the scurvy and to enable them to commence the permanent works." Thus was set the pattern of activity that would define the early years of the post, in which soldiers became builders and farmers as much as, or perhaps more than, infantrymen. Yet Leavenworth still had intentions of constructing a major fort. Indeed, the War Department decreed that the fort at the St. Peter's should be an impressive structure, large enough to house as entire regiment. The sparseness of good building timber remained a challenge, however. Prairie covered most of the territory around Bdote. Woodlands were confined mostly to river bottoms and lakeshores, and much of this was soft woods such as cottonwood, not useful for construction. To remedy this situation, Colonel Leavenworth sent a party to survey the area of the Mississippi above St. Anthony Falls, particularly the Rum River, where he was told there were stands of good timber. Meanwhile, his officers determined that a sawmill could be constructed near the falls. During the following winter, the first of many work details was sent to the Rum River area to cut white pine logs to be floated down the Mississippi during the spring thaw.[18]

In August 1820, Colonel Leavenworth began negotiating a new treaty with the Dakota concerning the military reserve—the land claimed for the use of the military at the confluence. He was authorized to do so as, in the absence of an Indian agent, the post commander could act as the agent. His motives are not entirely clear, however. Perhaps he sought to clarify the Pike treaty or add to it. The agreement Leavenworth signed on August 9 was a curious one. Most of the lands described were actually south of the St. Peter's and Mississippi Rivers, and they did not seem to

Coldwater Spring, 2016. *Photo by Matt Schmitt.*

Colonel Josiah Snelling, about 1818. *Oil by anonymous artist, MNHS; gift of Mrs. L. W. Hall and Marion Snelling Hall.*

include the location on which the still-unnamed fort was being constructed, so he probably meant to add more lands to those in the Pike treaty. The document also included a grant of land to translator Duncan Campbell and gave all of "Pike Island" to Pelagie Faribault, the Dakota wife of fur trader Jean Baptiste Faribault. Whatever its intent, this treaty was never ratified, but its provisions, particularly the grants of land, continued to create confused and conflicted land claims.[19]

Henry Leavenworth would not be around to appreciate the fruits of this labor, however. In September, he was superseded in command of the 5th Regiment by the officer whose name the new post would eventually bear, Colonel Josiah Snelling. Like Pike and Leavenworth, Snelling was an officer from the "old army" mold. Born in 1783, the son of a prosperous Boston baker, he had risen

through the ranks of the Massachusetts militia to a regular army commission in 1808. He gained a reputation as a personally courageous and energetic officer, seeing action against Tecumseh's Indian Confederation at the Battle of Tippecanoe in 1811 and against the British in the War of 1812. It was during the Siege of Detroit in 1812 that Snelling, then a widower with a young son, met and married Abigail Hunt.

War was the best opportunity to earn distinction and promotion in the army of that day, and Josiah Snelling rose from a captain to brevet colonel in the brief span of two years. Following the war, he retained the rank of lieutenant colonel, serving as adjutant to the 6th Infantry, first at

Plattsburgh, New York, and after 1819 in St. Louis, where he served as lieutenant colonel in the 4th Infantry. From there, Snelling was promoted to colonel and assigned to command the 5th Infantry at its new post at the confluence of the St. Peter's and Mississippi Rivers.[20]

Snelling's actual appointment as colonel of the 5th Infantry dated from June 1, 1819, but he did not receive orders to join his new command until early 1820. Traveling from St. Louis was a slow process, further delayed by his orders to preside over a contentious court-martial between two 5th Regiment officers at Prairie du Chien. Writing to Secretary of War Calhoun in November 1820, Snelling gave a colorful assessment of the circumstances at the time of his arrival:

> The permanent barracks at this place are now so far advanced that I expect to remove three companies into them in a few days. I could not go into a narration of the causes which have led to the waste of so much time, and the consumption of a great quantity of materials to little purpose in erecting two sets of temporary barracks, without seriously implicating the judgement of my predecessor in command, it is only necessary for my own justification to say that on my arrival at Prairie du Chien in August, learning that nothing had been done, and finding that I was likely to be detained there some time by a Court Martial, I dispatched an express with an order to Col Leavenworth to collect materials for the new barracks. I arrived here on the 6th of Sept and on the 10th the workmen broke ground, as but two months have elapsed and we have a block of twelve rooms nearly completed, besides the cellar walls of a whole range of barracks & one block house stoned up & another range of cellars completely dug, I flatter myself Sir you will not think we have been dilatory.[21]

The army officer corps was a small community in the 1820s, and Leavenworth eventually learned of Snelling's less-than-flattering comments when they were published in the *National Intelligencer* newspaper. He complained to Quartermaster

General Thomas Jesup that Snelling aimed at his (Leavenworth's) "destruction" while working to "puff up his own purpose." He would not forget the slight. The report also shows that Colonel Snelling decided on the location and basic configuration of the fort fairly quickly upon his arrival in September 1820. The design of the fort was entirely Snelling's, in keeping with the army's practice at that time of leaving such matters to local commanders. The configuration, a diamond shape running roughly east to west, fit the shape of the prominence, and his decision to use stone was a consequence of both the lack of readily obtainable timber and the abundance of easily worked limestone available near the site. He decided to name the new post Fort St. Anthony, after the nearby falls.[22]

Despite the colonel's glowing report, actual construction went slowly, and the troops, along with accompanying civilians, continued to live in the temporary structures at Camp Coldwater and Cantonment New Hope for one or more winters. According to fur trader Philander Prescott, who had recently arrived with additional stores for the post sutler, "The troops passed the summer at Camp Coldwater, and in the fall moved back again to the old cantonment and passed the winter, and got out timber for the soldiers' barracks and before the autumn of 1823 nearly all the soldiers had been got into quarters and considerable work had been done on the officers' quarters. . . . The Indians were all peaceable, and all things progressed peaceably and with all the speed that was possible for soldiers (for there is no hurrying of soldiers—they go just so fast, and out of that pace you cannot drive them)." In fact, there were just not that many men to do the work, probably fewer than two hundred, given the losses to disease. They had other duties as well, including planting and tending several hundred acres of gardens and soldiering when they had the chance. Colonel Snelling also noted that many of the building supplies so laboriously hauled upriver in 1819, particularly nails, glass, and iron, were expended in building Cantonment New Hope and Camp Coldwater. And despite the new sawmill, acquiring lumber was a problem, particularly as there were not enough draft animals to move it. "For two years," Colonel Snelling wrote in 1824, "the number of horses and working cattle were by no means adequate to the purposes of building and agriculture; our timber was cut nearly ten miles from the garrison and some distance from the river and has generally been procured with immense labor."[23]

The entire project, in fact, required "immense labor." From the time of Colonel Snelling's arrival in September 1820 until the end of the following August, 140 different soldiers, a substantial part of the garrison, spent a total of 18,696 working days on tasks relating to building the fort. Some labored only a few days, but most worked 100 or more. In particular, thirty-five soldiers toiled over 200 days of the year, and seven put in more than 300 days. The total extra pay for these men, most of whom were privates, came to $2,804.40, a considerable sum in 1821 dollars. Not surprisingly, several of those working the longest hours had skills related to the task. William Goddard, for example, who enlisted in the 5th Regiment in 1818 at Detroit, was born in England and was a stonemason by trade; he worked 288 days on the fort. John White, 308 days, had been a brickmaker in Pennsylvania before enlisting in 1817. Jacob Strause, who was a cooper when he enlisted at Philadelphia in 1815, earned an extra $46.20 for the 308 days he labored on Fort St. Anthony.[24]

Additional building supplies were ordered as well. For the years 1821 and 1822, these included 653 pounds of nails, 200 feet of glass, 500 pounds of iron bars, 2 gallons of white lead, 1,292 pounds of lead, 5 gallons of linseed oil, 20 pounds of putty, 40 pounds of steel, 16 dozen screws, and 4.5 gallons of turpentine. Most of these materials were brought up the Mississippi from St. Louis via keelboats, but some came from more local sources, including iron and steel bars and nails purchased in November 1821 from Joseph Rolette, the American Fur Company agent at Prairie du Chien. Colonel Snelling also acquired additional work animals from local sources, as he explained in a letter to auditor Peter Hagner in March 1823:

> I have to inform you, that for the public service at this post a considerable number of horses and working cattle are required, they are employed in drawing wood for fuel, hay for our stock, stone, brick, lime, sand, and plank for building, and for all the purposes of agriculture. Previous to the purchase referred to, we had but three horses at the post, I have since . . . increased the number to twenty one horses and mules . . . and for all these we find full employment. . . . I was induced to purchase them here in preference to requiring them of the Quarter Master in St. Louis, because they could be provided much cheaper, and the risk of driving them through the Sac nation (notorious horse thieves) was avoided.[25]

Most of the animals appear to have been purchased from fur traders; Henry K. Ortley provided eight of the horses. Among the oxen acquired was an animal with an interesting history, detailed by Snelling: "This ox belonged to a drove purchased by Lt. Fields for the Council Bluffs [probably Fort Atkinson] in 1819 but strayed, after three years absence was taken up . . . and driven to Ft. St. Anthony." One way or another, the colonel made up his shortage of work animals.[26]

By August 1824, Colonel Snelling could point with pride to a very substantial set of structures built or under construction on the bluff above the rivers. "Many things have conspired to delay the completion of the buildings until this time," he wrote to Quartermaster General Jesup, "but I confidently believe that the present year will close our labours." His report noted the loss of materials, shortage of work animals, difficulty in obtaining timber, and "my misfortune to see the

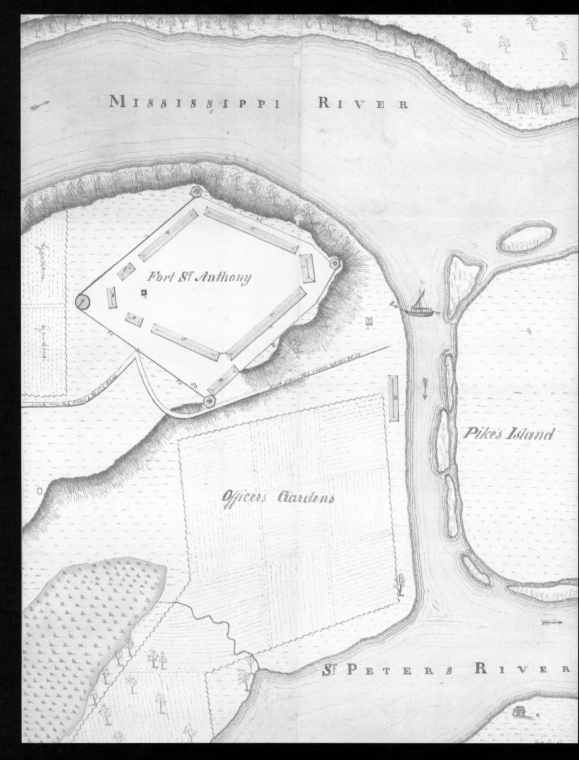

Topographical sketch of Fort St. Anthony drawn by Sergeant Joseph E. Heckle in 1823; Major Josiah Vose made the marginal notes, identifying various structures. These numbers do not correspond to Snelling's identifiers; Snelling may have been using a version of Heckle's map that has not been found. *MNHS*.

Colonel Snelling's

1824 report is worth quoting at length, since it describes not only the structures of the fort and the falls but also how they were adapted to the landscape. The numbers he mentions refer to his accompanying map, which is evidently similar to that prepared by Sergeant Joseph E. Heckle in 1823.

Fort St. Anthony is situated on a high point of land immediately at the junction of the river St. Peters with the Mississippi, its elevation is one hundred and ten feet from low water mark, it commands the channels of both rivers and the adjacent country within point blank distance; its peculiar form was chosen, to adapt to the shape of the ground on which it stands; on the north side the hill is a perpendicular bluff on the south the ascent is steep and a road has been cut by which stores, wood, etc. are conveyed from the landing to the garrison, between this and the St. Peters is a fine bottom containing about fourteen acres which is laid out in gardens, the whole work is surrounded by a stone wall, twelve feet high the arrangement of the several buildings and the uses, to which they are put, will be best understood by referring to the accompanying plan with the following explanations;

No. 1. The Commanding Officer's quarters is a stone building containing in the first story two large rooms and two bed rooms, with a spacious hall in the centre; in the basement, a kitchen and offices; it may be proper here to observe that all the officers quarters are one story high in front, with a basement story pointing to the rear, the level of the parade being higher than the natural elevation of the hill on the south side

No. 2. is a semicircular battery, the guns of which command the two rivers, and the two first angles of the curtains, as represented in the plan.

No. 3. The officers quarters built of wood, containing in the first story fourteen rooms with a small bed room annexed to each and in the basement a kitchen and pantry, six of these kitchens have cellars under the parade.

No. 4. The Commissaries and Quarter Master's store, this building is erected in a chasm of the hill, it forms a part of the South west curtain and is four stories high in front; its rear facing the interior of the garrison has the same elevation with the wall, it is a large and well finished stone building capable of containing four years supply of provisions and Quarter Master's stores.

No. 5. South battery, of stone: this building like the store presents a higher front towards the river than the garrison it has three stories, in the two lower, loop holes are opened for musketry, which command the landing the road and the St. Peters, on the upper story, cannon are mounted which flank the two curtains represented in the plan.

No. 6. Is a stone barrack of ten rooms, six of them are now occupied as a Hospital and divided as follows from the left, kitchen, sick ward, convalescent ward, store room, stewards room and surgery; the three rooms on the right are occupied by the suttler's family and one is vacant: this part of the buildings was originally designed for Company F which was removed to Fort Edwards 1822.

No. 7. is a stone building with a shed roof, attached to the wall, containing six work shops for Blacksmiths, Carpenters, Wheelwrights, etc. and a Bake house.

No. 8. The school house.

No. 9. The Guard house on the opposite side of the gate, and having the same exterior form: it contains a Guard room, a prison and two cells for solitary confinement.

No. 10. The Magazine.

No. 11. A stone tower, the platform for cannon elevated twenty two feet above the parade for the purpose of commanding the prairie in rear of the garrison; the two lower stories are furnished with loop holes for musketry.

No. 12. The suttler's store, corresponding in size and appearance with the school house opposite.

No. 13. A Well, twenty four feet deep, quarried through solid limestone, at that depth the water rushes in through a fissure in the rock in sufficient quantities to answer all the ordinary purposes of the garrison. It is uncommonly cold and clear.

No. 14. A stone barrack of ten rooms, corresponding with No. 6 apposite, occupied by the troops.

No. 15. North battery of stone, flanking the adjoining curtains as represented in the plan; this battery as well as the opposite one are erroneously called, Block houses by Lt. McCabe.

No. 16. Barracks of sixteen rooms, of wood occupied by the troops.

No. 17. The road, which has been constructed with immense labour; it was once entirely destroyed by a heavy storm and rebuilt with an exterior wall, sixteen feet wide at its base.

In addition to the above described buildings is a stable capable of containing one hundred horses, yet unfinished; the buildings in the garrison are all completed, excepting three rooms in the officers quarters, and the bake house; the old bake house, built in 1819 is at the foot of the hill and is still occupied.

The buildings at the Falls of St. Anthony are a Grist Mill, a saw mill, and two barracks containing five rooms. The Grist Mill is a fine building of stone, twenty five feet high from the bottom of the cog-pit to the eaves, it is furnished with an excellent pair of burr stones and the flour manufactured the last year was equal to any in the world. The saw mill is a wood frame building in the ordinary form, the machinery is of the best kind, we have sawed in twenty four hours, three thousand five hundred feet of pine plank; both of these mills are supplied with water from the falls, it is taken out at the table rock and conducted to the mills by a race placed on the right bank of the river.

The barracks were built for the accommodations of the workmen, but they are large enough to contain a company, should the conduct of the Indians ever render a guard necessary, the stone mill is a citadel of itself.[i] ᶿ corps [the 5th Infantry], two seasons in succession reduced to a mere skeleton by discharges [men being released from service at the end of their enlistment]." Snelling also commented that two of the new buildings were wooden, though stone was more plentiful, "but at the time they were erected we had but two or three masons and a large number of carpenters which determined me to give wood the preference."[27]

For its time and place, Fort St. Anthony represented the largest complex of stone structures for hundreds of miles around. It also contained the largest community of European Americans west of Prairie du Chien, and the largest between that outpost and the British colony founded in 1811 by Lord Selkirk near Fort Gary on the Red River. So the 5th Regiment's new post was quite isolated from other European American communities, a US colony plunked down in the land of the Dakota, hundreds of miles from the nearest source of supply or reinforcement. General Edmund P. Gaines, Colonel Snelling's superior officer and the commander of the army's Western Division, made an inspection tour in 1824 and was so impressed by the new post that he recommended the name be changed from Fort St. Anthony to Fort Snelling, in recognition of the work and dedication of the officer commanding.[28]

Although the 5th Regiment was headquartered at Fort Snelling, it had detachments of troops occupying Fort Crawford at Prairie du Chien and Fort Armstrong at Rock Island. Army organization for most of the nineteenth century was based on the regiment, which was divided into ten companies designated as A to K (omitting the letter J because it resembled I in handwriting of the era). The size of a company varied over time, but it could be from fifty to one hundred soldiers, though circumstances could leave them much smaller in number. Companies were commanded by a captain, who was assisted by a lieutenant and one or more second lieutenants and a number of noncommissioned officers. When occupying more than one post, a regiment's commander sent several companies to each, and the longest-serving captain among the several companies would be designated as post commander and sometimes given the "brevet" or temporary rank of major while in charge.

What did the Dakota make of all this activity in their territory? There is little information on their reaction to such an elaborate fortification going up at Bdote. Indian Agent Taliaferro gave accounts of the various chiefs who visited him for their share of promised gifts, some of them even wearing American medals. With the fort and the presence of the garrison, however, came change. As demands grew for use of local resources, particularly wood and game, the Dakota increasingly found their world shifting. By the end of the 1820s, it would be "altogether changed," in the words of the leader of Black Dog Village.[29]

The deployment of soldiers and delivery of supplies was an ongoing challenge, given the type of transportation used and the character of the Upper Mississippi. Steamboats were coming into use by 1820, but they were relatively underpowered and primitive, and they required deep water. The principal way to move large cargoes on the river at that time was either by flatboat—essentially a large, flat-bottomed barge that could only drift downstream—or by keelboat. The latter were crafts fifty to eighty feet long with fixed keels. They were usually propelled by deckhands pushing long poles along the riverbed, but most also carried a square sail that could be used with a favorable wind. Though cumbersome, they had the advantage of being able to work upstream and could carry several tons of cargo. Keelboats were the preferred means of trade and travel on much of the Ohio and Mississippi Rivers, and they were also used by army expeditions such as Lewis and Clark's Corps of Discovery and Pike's 1805 expedition. Yet their deeper draft created problems in areas of shallow water, and it was the variations in water level that made navigation of the Mississippi above St. Louis most challenging. No aids to navigation existed here, not even accurate maps. Boatmen knew their location by memorizing landmarks along the river, and they learned to read the surface of the water to detect snags and sandbars. In certain seasons, particularly late summer and early autumn, the river above Lake Pepin could be very low indeed. In times of drought, some places were so shallow they could be waded across. In mid- or late November, the river froze north of Prairie du Chien, and it did not open again until sometime in April. Even then, in years of large snowmelt and heavy winter rains, high water could delay the arrival of boats at the confluence until May.[30]

With this sort of irregular communication and supply, the post had to be as self-sufficient as possible. So along with its buildings, walls, and mills, the fort developed a considerable farming complex. Items supplied to Fort Snelling in the early 1820s included scythes and sharpening stones for them, as well as plows, hoes, and spades. Horses and oxen used in building the fort were also employed in the garrison's ambitious farming operations. Colonel Snelling reported in 1823, "We have a hundred acres sowed with wheat, and shall probably plant about the same quantity in corn; if the yield is as great as last year no contract for flour will be necessary."[31]

As with many of the colonel's reports, his early impressions turned out to be overly optimistic. A few more seasons of experience with agriculture in the Upper Mississippi Valley tempered his views. In 1826, he complained to Commissary General of Subsistence George Gibson, "I am weary of crying out, the blackbirds, the blackbirds but it is nevertheless true in a few days they destroy the labour of months. . . . Wherever they alight, the corn, wheat and oats immediately disappear. . . . My experience of six years has taught me to place

no dependence on our crops for a supply of bread stuffs." Indeed, the army's experience in farming at Fort Snelling mirrored what the first white farmers in the area would discover: this land could be bountiful, but good seasons might be quickly followed by disastrous ones, and the risks were considerable. By the end of the 1820s, however, the soldier-farmers seem to have adjusted their efforts away from cereal crops and enjoyed more success. Captain James Gale, the post commander in 1829, could report to Commissary General Jesup, "There is raised the present year upwards of 4000 bushels of potatoes, 800 bushels Rutabaga, 400 bushels of common round turnip, 4000 bushels of corn and upwards of 10,000 head of cabbage. There is about 60 tons of hay cut . . . for the quarter master dept., twenty tons of which is under cover the remainder in stacks." An impressive supply for the two hundred or so souls wintering at the fort that year.[32]

In addition to growing crops, the frontier posts were expected to raise a certain amount of their own beef and pork. To supplement the initial supply of draft animals and a few milk cows, the army endeavored to provide the new fort with beef cattle. This ended up as a confused affair illustrative of the difficulties of communication between the St. Peter's post and supply officers in distant St. Louis. Joseph Rolette, the agent at Prairie du Chien, offered to supply beef "on the hoof" for the new garrison from "local" sources (or more local than St. Louis). The officers in St. Louis and the supply officer for the 5th Regiment seemed to miss communications. Rolette's proposal was rejected after he had procured the cattle, while the St. Louis officers sent up a "drove of beeves," few of which made it as far as the mouth of the St. Peter's River. Writing again to General Gibson, Colonel Snelling vented his frustration.

Sir, it is utterly impossible to subsist cattle driven here in Oct. and Nov. without previous notice, the season for procuring subsistence has then gone by; a sufficient quantity of hay had been cut for the working cattle, cows, and a few Beeves we intended to keep for slaughter through the winter and no more; at the time of their arrival the Prairies were dry, and in many places burnt over, under these circumstances all of the disposable force I have could not prevent their straying, or even if it could have been prevented, they must have starved.[33]

As with the problems relating to crops, the issues with livestock would be overcome to some degree. The post would often furnish its own tallow for candles and even ship surplus hides on occasion to the commissary at St. Louis, but its soldiers never had the same degree of success with animals as with gardens. Colonel Snelling later reflected on this fact: "As for raising stock I was always of the opinion that it could be done but not while our commissaries were exclusively appointed from West Point. Some of them are men of science and all can tell you the price of wings, swords and sashes, but none how long a cow goes with calf, when it should be weaned, or how stock should be wintered, it forms no part of their education and in fact they think themselves above it."[34]

⁓

For soldiers of the garrison, army life must have seemed a fixed reality, a world governed by the daily routines and duties found on most posts. For the next few decades, as a community of traders and other civilians (discussed in the next chapter) grew up outside the walls of the fort, the everyday round of reveille, roll call, quarters, guard mount, sick call, meals, and fatigue duties—carried out to the signal of bugle, fife, and drum—ruled the lives of those who served at Fort Snelling. Musicians also played at funerals for both soldiers and civilians. For example, Lawrence Taliaferro noted a "band of music beating a solum [*sic*] dirge" for the funeral of young Melancthon Snelling Green, the son of Lieutenant Green.[35]

Farming on a large scale and putting up hay were important duties in spring and summer; gathering wood—essential for heat, cooking, and making the charcoal used by the blacksmiths—was a year-round activity. Anywhere from 800 to 1,400 cords of wood were gathered, cut, and consumed in any given year.[36]

Finding and transporting all of that fuel presented challenges, particularly in a landscape dominated by prairie and regularly set alight by its inhabitants. "The Prairies are generally on fire at this date," Lawrence Taliaferro observed in October 1827, "every year at this season—the moment they start upon their winter or fall hunt, [the Dakota] set fire to the prairies, which starts off the game in every direction, to avoid otherwise certain destruction."[37]

Wood details, or "parties," were usually made up of three to fifteen men commanded by a non-commissioned officer. Using a team of horses and a wagon, they ranged as far as ten miles from the fort in search of usable firewood. Their work could leave them out overnight, occasionally resulting in mischief of one kind or another. Men sometimes deserted the group in search of whiskey, and on other occasions harassed Dakota women. Such treatment could range from making unwanted approaches to rape. This became an ongoing problem that only worsened over time. Samuel Pond noted that by the mid-1830s Dakota women avoided places "where they were likely to meet one of our soldiers without an escort to protect them from insult." Taliaferro would struggle to address these issues.[38]

Other detached duties could take soldiers even farther afield. When new buildings or extensive repairs were needed at Fort Snelling, a detail was sent into the white pine regions along the Rum River to cut logs for the government sawmill at St. Anthony Falls. There does not seem to have been a formal agreement with the Ojibwe for the army to cut wood in their territory. This practice was infrequent, however, and the impact relatively small, so there may have been no need for legalities. These expeditions were led by an officer assisted by a sergeant and could last for several months. Over the winter, trees were cut, trimmed, and hauled to the banks of the Rum River. In spring, rafts of logs were floated down to the Mississippi and thence to the Falls of St. Anthony, much in the way commercial loggers would later operate. It was an inexact method of transportation, however, and sometimes other parties of soldiers were sent forth to scour the riverbanks for missing logs.[39]

Particularly trusted soldiers were sent downriver to Prairie du Chien to bring back the post "express" or mail. In the summer months, this all-important communication link to the outside world might come by steamboat or keelboat as part of a regular supply run, though soldiers sometimes made the five-hundred-mile round trip by canoe. In winter, with the Mississippi frozen over, a small party would literally risk life and limb to make the journey by dog team. Or couriers would start from Fort Crawford and Fort Snelling on a specified date with the understanding they would meet at Wabasha's village, a convenient halfway point, to exchange mailbags. In the late 1830s, this necessary but perilous mail run was taken on by contractors, often experienced fur traders.[40]

Life could be harsh for the fort's soldiers even by nineteenth-century standards. Isolated from eastern cities and towns, subject to stiff, sometimes severe discipline, they toiled as farmhands or construction laborers while living in a climate more extreme than most of them had ever experienced. During bitterly cold winters, they slept four men to a bunk, two men sharing each "box," under layers of blankets and "hay sacks," leaving this relative warmth for their turn at guard duty. In severe subzero weather, guards might

A note on Fort Snelling's post office register for January 1838 demonstrates the challenges of regular delivery: "This mail was due in Dec. 1837 on the old contract, it was delayed by the Ice range in the Mississippi which rendered it impossible to cross." *Henry H. Sibley Papers, MNHS.*

change every quarter hour, but frostbite was still common.[41]

Other factors compensated somewhat for these disadvantages of frontier army life. Soldiers from American and European laboring classes were accustomed to hard work in harsh conditions—and to hunger. So regular meals were a significant consideration. In 1837, a typical daily ration included three-quarter pound of salt pork and eighteen ounces of bread or superfine flour; twice a week, "at the discretion of the commanding officer," each soldier might receive a pound and a quarter of fresh beef. In addition, every one hundred rations, approximately a company, received a biweekly ration of two quarts of salt, four quarts of vinegar, eight quarts of beans or peas, four pounds of coffee, and eight pounds of sugar. The post gardens contributed an important supplement to this essentially bread-and-meat diet, providing root vegetables like carrots, onions, potatoes, beets, and rutabagas, as well as great quantities of that nineteenth-century standard, cabbage. Other foodstuffs, such as dried fruit, almonds, butter, cheese, eggs, sausage, dried mackerel, and spices, might be found, for a price, at the sutler's store. Enlisted men typically ate together as a company mess, with cooking duties rotating among the soldiers, though cooks could be hired as well. Except in periods of extreme weather, soldiers ate regularly and well.[42]

Regardless of where they were stationed, soldiers were expected to be clean. This meant a standard of personal hygiene probably greater than that of working-class civilians, involving occasional bathing, for example. Even soldiers' clothing had to pass inspection.

To help ensure its standards could be met, the army provided for an official civilian female presence at each garrison: laundresses. Under an 1802 regulation, every post was allowed one laundress per seventeen men, but no more than four per company. The job was an employment opportunity for wives of enlisted soldiers, and because their names were not included on post records, little is known about them. Army regulations required that laundresses wash three times a week in the summer and twice a week in the winter—hard, heavy work for a group of soldiers, officers, and/or dependents assigned to each laundress. In addition to a daily food ration, the women received bedding (a straw mattress) and were entitled to medical treatment from the post surgeon. They were also guaranteed wages, with the rates set by the local Council of Administration, a group of officers that regulated the sutler's fees and purchased books and periodicals for the post library. If their rates matched those paid at Fort Atkinson in the 1830s and 1840s, a washerwoman would receive fifty cents per month for each soldier, two dollars for each officer, and four dollars for married officers (with extra pay for each dependent). A woman's income fluctuated with the number of soldiers she was assigned, and it would be difficult to assess how far her income stretched. Though they were not "ladies," the great majority of them were respectable members of the community. At Fort Snelling, the laundresses worked in the washhouse located on the riverbank below the fort; on at least one occasion, it was swept away in a particularly bad spring flood.[43]

⌒

The men of the garrison also had access to a physician, the post surgeon. Fort Snelling's regularly assigned assistant surgeon treated sick or wounded soldiers. Though the medical practice of the era emphasized bleeding and purges, which sometimes did patients more harm than good, other basic treatments worked reasonably well. For instance, post surgeons made extensive use of quinine in treating fevers, particularly an illness described as "intermittent fever"—probably malaria.

Still, the constant labor and climate at Fort Snelling took their toll, as records from later decades show. Between 1838 and 1850, some twenty-two soldiers died, according to the surgeon's reports. Some men succumbed to bronchitis or diarrhea, common causes of death in those years, while others died from infections such as typhoid fever, as in the case of an unfortunate soldier who stepped on a nail. Alcohol consumption, a problem well known to the surgeons, also contributed to several deaths, including one man who became drunk and fell from a cliff. Yet the loss of twenty-two men over twelve years of reporting was likely lower than the average mortality rate. Death records for the civilian population prior to 1850 are scarce, but statistical studies suggest white adult males in 1850 had a life expectancy of forty-six to forty-eight years. Even within the army, several of the post surgeons note a lower incidence of disease at Fort Snelling; spikes in illnesses arose not from local sources but were diseases, like malaria and measles, that troops or recruits brought with them from other locations. While mortality was certainly a factor, more men were lost to the garrison through being discharged for disability, either disease or injury. Inguinal hernias, often caused by hard work and the strain of heavy lifting, were a frequent cause of disability at the fort; they were debilitating, painful, and essentially untreatable at that time. Chronic diarrhea, epilepsy, problems with vision, and eye infections were also listed as reasons for disability discharge.[44]

In the course of a typical quarter, a surgeon might see 100 to 150 cases, or 30 to 50 men per month, out of a garrison of 80 to 150 soldiers. Most of these were for minor issues, and the men returned to duty, but the post hospital, built in the 1820s, always had eight to ten patients. The surgeons were also called upon to treat soldiers' and officers' families, and the Indian agent might ask them to treat nearby Dakota or Ojibwe. Some found the work overwhelming. Dr. E. Purcell reported in 1824 to the surgeon general, "The duties to be performed among the soldiery at this post does hardly constitute one half of the labor, that among the officers their wives, children; soldiers wives & children together with the general attention that must be paid to the police [cleaning] of the Hospital etc., make the duties so fatiguing that I find it necessary to ask for an assistant."[45]

Even finding soldiers who could act as hospital stewards could be difficult: "I was necessitated to perform the minor and to me more harassing duties of hospital steward. It will hardly be credited; but the fact is not the less true, that there cannot be found in this Reg't. [5th Infantry] a single noncommissioned officer or private possessing sufficient sobriety, capacity and honesty to fill a station of such comparative irresponsibility. I have therefore no other alternative than to continue to enact the part and so extensive is the empire of drunkenness that until other troops arrive I need hope for no relief."[46]

Even when a competent steward was found, the frequent turnover of troops made keeping him on the post difficult. In 1841, surgeon George Turner had an ideal hospital steward, a Swedish immigrant named Hugo Ferdinand: "He is by birth a Swede, but speaks our language well, is 24 years of age and evidently a man of considerable education, possessing besides other more important information a good knowledge of Latin, which will facilitate him in the study of Pharmacy in which he has already made very considerable proficiency." But despite Turner's repeated efforts, this paragon was transferred away and left the service in 1844.[47]

Fort Snelling as completed contained and sheltered the world of army life and routine. Most of the post's mission and purpose, however, lay outside its walls, with the Dakota, the traders, and the Indian agency intended to regulate them.

Fort Snelling, about 1844. *Oil by John Casper Wild, MNHS.*

4. The Agency and Its Community

THE WALLS OF FORT SNELLING IN THE 1820S seemed to create a clear demarcation between inside and outside. But the military garrison, tending to its mills and gardens, was part of the community growing up around the fort.

Dakota people of the region continued to visit sacred sites in the Bdote area. But they had another reason to call at the fort. About a quarter mile from its gate stood the buildings of the St. Peter's Indian Agency. These consisted of a dwelling for the Indian agent and later his assistant, plus a Council House where official meetings with Dakota, Ojibwe, and in later years Ho-Chunk leaders took place. There also stood several storehouses that held the gifts (manufactured goods like glass beads, colorful fabrics, and knives), flour, salt pork, gunpowder, lead, and small quantities of whiskey that were dispensed by the agent. All of these buildings were constructed of wood that had been cut on the Rum River, sawn into planks at the army's mill, and—like the fort—built with the labor of soldiers. The army was not simply being generous to the Indian Department. The agency was at the center of the fort's mission and a major reason for its existence. Although the post commander sometimes acted as Indian agent when the appointed agent was absent or unavailable, the task of regulating trade and promoting peace between the tribes fell to the Indian agent.

The agent, with the support of the troops, was an agent of the US president, expected to impose an American system of regulation and rule of law upon Dakota and Ojibwe people whose lives for centuries had been governed by their own successful cultures and, above all, personal relationships. The agent's duties would eventually include establishing and enforcing boundaries between Indigenous nations, which would both prevent disruption of the fur trade and allow the United States to more easily accomplish land cessions. Some form of exchange had been going on between Europeans and Native Americans since the sixteenth century, and for much of that time, because of the remoteness of European power, the relations had been on nearly equal terms. French and later English traders depended on developing kinship relations, through marriage or adoption, to become part of the community—and thus to be able to obtain furs. They were obliged to respect Native cultures, having little or no direct means to impose their will by military or economic force. Even the consolidation of trade by the British fur companies in the later eighteenth century had its limitations, and maintaining a level of kinship remained important. Powerful figures in the trade like Robert Dickson had married Dakota or Ojibwe women and relied on their relations to help further their influence.

But the arrival of American troops stationed in substantial, permanent posts changed the situation. The approach of the American government was also totally different from its European predecessors. Americans meant not only to control the trade but to change and ultimately disinherit the Indians as well. And although their stated objectives were "enlightened" and "benign," Thomas Jefferson's 1803 letter to William Henry Harrison showed their ultimate aim: to separate the Native people from their ancestral lands. Once opened to colonization by white European Americans, the land they saw as empty wilderness could be made fruitful and profitable.[1]

The duties of Indian agents were defined by the Indian Trade and Intercourse Act of 1793 and its various amended versions. Agents were to control trade by issuing to traders licenses that specified where they could establish trading posts. Theoretically, they could limit the number of traders operating in a region and bar or expel any who violated the trade laws. Following the War of 1812, agents were specifically ordered to suppress foreign traders, especially those from British Canada. In addition to regulating trade, the agents were to encourage the "civilization" of their Indian charges by introducing and promoting European American-style agriculture and education—and further, to ensure peace among the tribes. To enforce these policies, the agents were to rely on frontier garrisons, like the troops at Fort Snelling. Much depended on the cooperation between post commanders and Indian agents.[2]

Agents were political appointees, often former

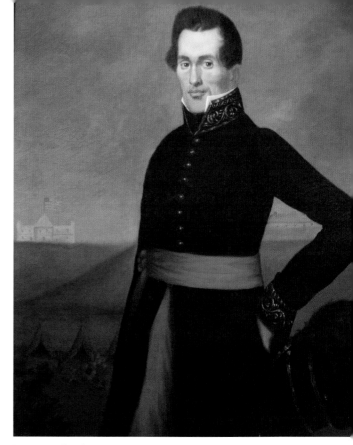

Lawrence Taliaferro, Indian agent at Fort Snelling, about 1830. *Oil by anonymous artist, MNHS; gift of Mrs. Virginia Bonner Pesch and John F. Bonner.*

army officers, traders, or well-connected politicians. Major Lawrence Taliaferro, the first agent at Fort Snelling, was an unusually honest, honorable, and occasionally difficult man, with a strong sense of duty and a clear vision of his mission. He was also a slaveholder who rented his human chattel to officers at the fort. Born into a prominent Virginia family, he received an excellent education, and upon enlisting in the army during the War of 1812, he was quickly commissioned as an ensign, serving at Fort Belle Fontaine on the Missouri River near St. Louis and later as a lieutenant at Detroit and Chicago. But Taliaferro's promising career as an officer on the frontier was cut short by illness. As he convalesced at his family's home in Virginia in 1819, his connections and experience in the West led him to a new opportunity. President James Monroe was looking to appoint an Indian agent for the proposed post at

Lawrence Taliaferro

also had an influence, one that he may not have foreseen, on how the early history of Fort Snelling has been told. The agent kept a daily record of his activities and interactions with Native peoples, soldiers, and civilians. This journal included extensive quotations from speeches given by both Ojibwe and Dakota leaders, which seem to be verbatim but are in fact translations provided by Taliaferro's interpreters, Scott and Duncan Campbell. Though these men were of Dakota ancestry and understood the language and its nuances well, the differences between Native languages and English made literal translation between them problematic. There is no way to know to what extent the Campbells tailored their interpretations to please their boss—or, conversely, softened Taliaferro's often paternalistic tone when translating the agent's words to Dakota leaders. Further, it's unclear from the journals whether Taliaferro's accounts are based on transcriptions or his own memory of a given speech. So while the agent's journals are among the few written, contemporary sources documenting Dakota and Ojibwe perspectives, they also reflect problems in translating languages and cultures and, to some degree, his own views and prejudices.

the confluence of the Mississippi and St. Peter's Rivers, which lay just south of the loosely defined territory that was shared by the Dakota with their neighbors to the north, the Ojibwe. President Monroe chose to offer the position to Lieutenant Lawrence Taliaferro.[3]

The offer was irresistible, as Taliaferro would note in his journal: "I left the army with extreme reluctance But President Monroe, having selected me from the line of the Army proposed the station upon me in a manner so kind and confirming that I was forced to yield—He had heard a good report of me—That I was above my rank—promotion too slow—my health not good—would have more command of my time—could return to the army at some future date—my pay was more than doubled." Most traders and some army officers would come to despise Taliaferro, while many of the Dakota and Ojibwe leaders would regard him with respect and even affection. By any measure, Lawrence Taliaferro had a powerful influence on life around Fort Snelling in its early years.[4]

Conflicts with the traders over the enforcement of the intercourse acts formed a theme during Taliaferro's nearly twenty years at the confluence. Initially, he strove to replace the long-standing loyalty of the Ojibwe and Dakota to the British with their new "Father," the American government, personified by the president and his agent. This task was not easy, as many of the British traders had well-established relationships with tribal leaders, and none were anxious to change their ways. Taliaferro soon discovered that American traders in general had a bad reputation among the Dakota. In 1822, he complained to Secretary of War Calhoun about "the ill repute in which Americans stand with Indians as traders and the decided preference given by them to every thing British."

Taliaferro blamed trader Joseph Rolette for the situation, and with him the American Fur Company (AFC). Formed in 1818, the company had created a vertical monopoly that controlled every aspect of the fur trade in American territory, from acquiring and transporting trade goods to the marketing of furs in the United States and Europe. Many of its agents, like Rolette and his associate Alexis Bailly, had long experience in the trade; some had even worked for British trading firms. They resented Taliaferro's insistence on enforcing the intercourse laws, particularly regarding the trading of whiskey and other "spiritous liquors," which the traders saw as vital to their influence with the Indians. Earlier the agent

The St. Peter's Indian Agency seal used by Lawrence Taliaferro. *MNHS*.

had noted regarding the AFC, "They have been known to say, we defy you to take our goods from us or interfere. If you do our whiskey will be given to the Indians to defend our profits." Indeed, they viewed men like Taliaferro as "misguided philanthropists," ignorant of the realities of commerce and life on the frontier. In turn, the haughty Virginian saw the AFC and its agents as ruthless semi-barbarians, "Mississippi Demi-civilized Canadian mongrel English American citizens," whose sole object was to thwart the government's policies for their own profit—and whiskey was the primary means to that end.[5]

Another great challenge for Taliaferro came from the garrison of Fort Snelling itself. As a former officer, he generally regarded army officers as gentlemen, but he held no such illusions regarding the enlisted rank and file. Enlistees in the first half of the nineteenth century tended to be less-than-ideal soldiers. The army was both underfunded and undersized for its role on the frontier, and it was further hampered by the need to recruit men in an economy that offered

better opportunities, particularly to those born in the United States with the means and the necessary connections or capital to get a start. Low pay, strict and sometimes brutal discipline, and the possibility of being stationed far away from friends and loved ones were not attractions for most young men of the time. While some recruits may have had romantic notions of adventure in the West, it is likely that many were men looking to escape a difficult financial or personal situation, or they were newly arrived immigrants for whom the army offered both immediate employment and free transportation to the frontier and its perceived opportunities. Recruiting officers were not overly particular about a man's past or his habits. These men seldom had any experience of the frontier or firsthand knowledge of Native people, and they carried the racial and ethnic stereotypes held by white society in general, which regarded Indigenous people as racial inferiors living in an inferior, "savage" culture.[6]

Though early accounts do not record problems in relations between soldiers and Dakota people near the fort, by the later 1820s there were difficulties. While soldiers were on duty within the walls of the post or under direct supervision of officers, army discipline tended to prevent conflict, but the behavior of men on detached duty away from the garrison was another matter entirely. The nearly continuous wood-gathering details and cattle grazing became a particular source of conflict, with soldiers not too particular about where they found wood and fodder or how they treated the Dakota, especially Dakota women, whom they encountered. For example, in February 1826, Little Crow complained to the Indian agent that soldiers had burnt two of his bark lodges the previous autumn and "that soldiers who attend the cattle near his village, while his women were in search of their corn, abused them very much, one of the soldiers shot at one of the women and lodged two shot in her breast." The chief also protested that soldiers were cutting down the trees his people used to get bark for their summer lodges. Some men became so troublesome that the Dakota threatened reprisals: "Russick of F Company is much complained of, he having endeavored to force two or three different women who felt disposed to resist his embraces. The Indians have declared that unless this man ceases to interfere with their wives and daughters that they will be compelled to kill Russick in self defense."[7]

How did the Mdewakanton and Wahpekute find themselves in this situation? For decades, the Dakota had wished for closer access to traders and their goods and for the payment that Zebulon Pike had promised them in 1805 in exchange for the use of the land on which the fort stood.

The Dakota village of Kaposia, led by a series of hereditary leaders named Little Crow, on the site of what is now the city of St. Paul, 1840s. *Watercolor by Seth Eastman, MNHS*.

Alexis Bailly, 1858. *Oil by Theophile Hamel, MNHS; gift of the Charles P. Bailly estate.*

men of mixed European and Indigenous ancestry, fewer and fewer traders actually lived among the villages, instead occupying seasonal or permanent trading posts. Men like Alexis Bailly, of mixed ancestry yet educated in the ways of capitalism, consistently put business before kinship.

The trade itself was beginning to run into limitations of its own making. Credit and debt were its lifeblood. The AFC sold supplies to its various branches or "outfits" on credit. The outfits in turn furnished these goods, on credit, to its traders. They then supplied the various Indigenous men with whom they worked the traps, spears, guns, and ammunition necessary to hunt and trap the fur-bearing animals, as well as other items needed by their families, again on credit. The Natives killed the animals and dressed their hides, then brought the pelts to the trader, who wrote them off against the original debt. Ideally, the Native family would have enough furs to create a surplus in the account allowing the purchase of other goods; if not, the hunter could get them on credit, further increasing his debt.

The Dakota and other Native Americans involved in the fur trade had become like the employees of an extractive industry buying from the company store. In the years to come, it would no longer be a "trade" in the sense of bartering between equally positioned parties. The presence of the army meant the Dakota and Ojibwe could not expel the traders, even if they wanted to. Thus, the company was free to determine the price in furs that the Indigenous hunters paid for the goods. As long as the value of furs remained high, the system could work and the credit continued to flow through the organization. But when the value of furs began to fall, the relationship would change.[8]

By the later 1820s, large game, like deer, elk, and buffalo, were becoming scarce in the region, and the Mdewakanton were more reliant on fish, small game, and their crops. For those living near

When they made their agreement with Pike, they likely did not anticipate a large stone fort in their midst, or the violent, threatening actions of some of its soldier residents. Relations with the fort and the traders created a conundrum that would only increase over time, particularly for the Mdewakanton, who lived closest to the fort. Trade with the French and with traders licensed by the British had been based on personal relationships in line with their traditions. But the rise of the American Fur Company through the 1820s and the presence of the fort began to radically change these arrangements. As independent traders were pushed out of business, the personal kinship ties between traders and the Dakota began to unravel. Though the traders were still mostly the fort, at least, the occasional distribution of gifts and rations was becoming more important. It was this relationship with the agent that they did not wish to break. Further, war with the soldiers would be a risky venture. Though the Fort Snelling garrison was small, the Dakota knew that the Americans were powerful, and there was also the risk that other tribes—the Ojibwe and the Sac and Fox in particular—might side with newcomers against the Dakota. In this uncertain and shifting situation, war was a last resort, whatever the provocation.[9]

The Indian agency also worked to smooth over relations. Taliaferro repeatedly complained of soldiers' behavior to Colonel Snelling and his successors, urging them to stop the abuse. Additionally, he was quick to provide gifts, sometimes substantial ones, to encourage good relations. These presents could include considerable quantities of tobacco, gunpowder, lead, flints, cloth, tin kettles, blankets, and even, to the chagrin of the American Fur Company, whiskey. For example, in 1822, Taliaferro listed the items he distributed to Kiŋyaŋ, a prominent Wahpeton leader:

100 lbs of tobacco
50 lbs of powder [gunpowder]
120 lbs of lead [for making musket balls]
½ piece of selempore [a type of colorful, inexpensive cloth from India]
100 flints [also for muskets]
2 pairs of 2½ point blankets ["points" denoted the thickness of the wool]
1 pair 2 point blankets[10]

The agent also noted, "This Indian wears a 2nd sized Monroe medal. He is to receive a flag." The flag and the medal, which had an embossed likeness of President James Monroe, were tokens of Kiŋyaŋ's relationship with the American "father." Sometimes Indian leaders would exchange their old medals of King George III and British flags for the new American versions. While some of the smaller gifts listed above—the cloth and blankets, for example—were probably intended for the chief or his family, items like the lead, powder, and tobacco were clearly intended for the leader to distribute among his people, confirming his own generosity and status. Similarly, gifts of whiskey, usually small kegs amounting to eight or nine gallons, were distributed at feasts or ceremonies. Such generosity from the agent probably rankled the fur traders, who regularly exchanged such items with the Dakota and Ojibwe for furs.

—

The American soldiers, officers, and dependents occupying the new post at Fort St. Anthony must have felt as if they had arrived in a distant, foreign land. Not only did the region around the post seem wild and isolated, but the people of the area also lived in an utterly different kind of society from that of the United States. Here, two centuries of the fur trade had given rise to a racial and cultural amalgamation unknown in the "settled" regions of North America. French was more likely to be heard than English. Racial identity was far more fluid at the confluence of the Mississippi and the St. Peter's Rivers than in the East and Southeast, where the slightest hint of Black or Native ancestry might place a person in the lowest ranks of society—or a life of slavery. But here, far removed from Anglo-American society, race played out differently. This was a land where, as historian Mary Wingerd notes, "a single individual could conceivably trace a lineage that included Scots, French, American, Dakota and Ojibwe forebears among his extended relations."[11]

The military, of course, brought its own way of organizing society to the confluence. The army carried strong elements of class division, and men of color were not allowed in its ranks after 1820. But perhaps the most influential bearers of European American culture at Bdote were the army's dependents, particularly the officers' wives. Determined to bring what they regarded as the light of civilization into what they viewed as an alien wilderness, they did their best to establish a version of a refined, even elegant, life for their families. As practitioners of the nineteenth-century "code of true womanhood" who held that women should be "pious, pure, domestic, and submissive," they particularly shunned the mixed-race wives and children of the traders. Not only was their "half-breed" racial background suspect, but their active participation in their husbands' business activities was deemed improper by the ladies of the post.[12]

In the 1820s, the most conspicuous among the women at Fort Snelling was Abigail Hunt Snelling, the colonel's wife. Born in 1797 into a military family, she was in many respects the quintessential army bride. Abigail's father, Major Thomas Hunt, had been an officer in the army, first during the Revolution and later on a number of frontier posts, including Mackinac and Fort Belle Fontaine. The family accompanied Major Hunt to these postings, and young Abigail experienced something of the rigors of frontier life from an early age. When her father died in 1808, she was sent to live with her older brother's family in Detroit. It was here in 1812, at the beginning of the war with Britain, that she was courted by the older, dashing Captain Josiah Snelling. The two were married just as the Siege of Detroit began. Following the war, Abigail and her husband moved to Plattsburg, New York, then to St. Louis. In 1819, when Josiah was promoted to colonel of the 5th Infantry and ordered to the regiment's new post being built at the confluence of the St. Peter's River, there was no question about Abigail's accompanying him.[13]

The slow upstream journey, made even more tedious in the heat of summer when swarms of mosquitoes plagued travelers on the Mississippi, must have been particularly difficult for Abigail,

Abigail Hunt Snelling, about 1818. *Oil painting by anonymous artist, MNHS; gift of Mrs. L. W. Hall and Marion Snelling Hall.*

as she was pregnant. Two months after arriving at her new home, she gave birth to a daughter, Elizabeth. She was not the only baby at Fort Snelling in its early years. Charlotte Ann Seymour Clark, the wife of Captain Nathan Clark, gave birth to her daughter in the summer of 1819, while the 5th Infantry was making its long journey from Detroit to the St. Peter's. The little girl was named "Charlotte Ouisconsin" for her mother and the place of her birth along the Wisconsin River. Indeed, "Mrs. Colonel Snelling" and Mrs. Clark would work together in a number of ways, particularly regarding education and religion. They organized an informal post school and Sunday school. The latter also included adults and encompassed a Sunday worship service taken from the Episcopal Church's Book of Common Prayer.[14]

Abigail Snelling, Charlotte Clark, and other officers' wives also collaborated on entertainments. They were able to do so, in part, because they had the financial means to purchase extra comfort and luxury items both for their own households and for special community events from the post sutler. Every army garrison had a designated merchant or sutler authorized by the service to sell goods not provided by the army. It might seem that each sutler had a monopoly, but theirs could be a risky business. Sutlers purchased and transported goods on their own, but they did not set their own prices. A designated group of officers known as the Council of Administration determined the sutler's profit and adjusted prices by a formula based on the size of the garrison. Because of the difficulty of transportation in the 1820s, post sutlers at Fort Snelling were not likely to make much profit. In addition, they competed with agents of the American Fur Company, particularly Alexis Bailly and his father-in-law, Jean Baptist Faribault, whose trading posts were conveniently located at Mendota, across the Minnesota River from the fort. These traders quite probably had better access to goods and their transportation through the company's offices in St. Louis, and the officers of the fort and their wives were active customers. Bailly's accounts for 1825-26 show they bought ladies' shoes, combs, stockings, skeins of silk, yards of lace, and other fancy fabrics, as well as food items like kegs of butter, sugar, tobacco, and the ubiquitous whiskey.[15]

Bailly, with the help of his superior Joseph Rolette, also sold riding horses, along with bridles, saddles, horseshoes, and corn to feed them, to the officers, notably Colonel Snelling and Indian Agent Taliaferro. So while the colonel and the agent were confronting Bailly over selling whiskey to the Indians, they were among the trader's best customers.[16]

Although their purchases allowed Abigail Snelling and other women at the fort to entertain with some style, none of their civilizing efforts would have gone very far without domestic labor. Some of these workers were the wives or relatives of enlisted men who helped look after children or do household chores. But among the servants at Fort Snelling from its beginning were people brought in bondage. The institution of slavery, long established in the original thirteen states, was illegal but regularly practiced—and occasionally contested—on the Upper Mississippi.

The Northwest Ordinance of 1787, which governed the creation of new territories and states in the region north of the Ohio River as far west as the Mississippi River, clearly outlawed slavery. But Fort Snelling was located just across the Mississippi River, on land that the United States claimed through the Louisiana Purchase—officially a federal enclave in Indian territory—and slavery was recognized and practiced in French and Spanish Louisiana before and after the purchase of the region by the United States. Following the War of 1812, the area around St. Louis flourished, and since slavery already existed there, many southern slaveholders purchased land in the new Missouri Territory, bringing still more enslaved people. When the territory sought to become a state, those opposed to slavery attempted to limit its expansion in the first of several compromises over the issue. The Missouri Compromise of 1820 allowed Missouri to enter the union as a slave state, while Maine came in as a free state, maintaining the balance of power between free and slave states in the US Senate. More importantly, the southern border of Missouri (36 degrees 30 minutes north) was to mark the boundary between slave and free for the future. All territories north of the line were to be free; all south of it could allow slavery. Thus any expansion in the Upper Mississippi region would be free.[17]

The military lived by its own regulations, though, and after 1818, US Army regulations compensated officers for keeping personal servants, free or enslaved. Each officer submitted a monthly pay voucher naming the servant in order to receive the extra pay, which ranged from 15 to 30 percent of base salary. Officers who hired servants passed the pay along; those who held enslaved servants could simply keep the extra cash.[18]

Remarkable research by Walt Bachman has tabulated the presence of enslaved people on monthly military pay vouchers. Slavery was far more common among regiments raised in the South, like the 1st Infantry, than in the North, like the 5th. Between 1819 and 1828, seven officers of more than fifty in the 5th Regiment serving at Fort Crawford and Fort Snelling—including Colonel Josiah Snelling—submitted vouchers for a total of ten enslaved servants. From 1828 to 1837, thirty-three of the thirty-eight officers of the 1st Infantry, garrisoned at the same posts, claimed payment for a slave for at least one month. (About seven officers were assigned to Fort Snelling at a time.) The 1st Infantry was commanded by future president Zachary Taylor, a prominent enslaver. Colonel Taylor's father was a major slave owner in Kentucky, and the colonel himself owned plantations in Mississippi and Louisiana, where more than one hundred people were enslaved. Though the absolute number of unfree servants at the post was never very large, their presence at Fort Snelling for over thirty years in a region that eventually became a free territory would come to have much wider legal and political implications for the nation. Further, the presence of enslaved people at Forts Snelling and Crawford reinforced slaveholding in the region, specifically for those involved in the fur trade.[19]

With Missouri's new status, St. Louis established a market for sales of enslaved people and provided a source for officers at Fort Snelling. In 1827, Colonel Snelling purchased a woman named Mary and her daughter Louisa in St. Louis and brought them to the fort, setting the example at the top. By the early 1830s, officers and other officials at the post, particularly those with families, were purchasing "servants" there. Fur trader Alexis Bailly purchased an enslaved woman from an officer in 1831 and sold a woman, probably the same person, to his father-in-law, Jean Baptist Faribault. Lawrence Taliaferro brought three enslaved people with him to the confluence, and he sometimes leased them to officers. As the son of a slaveholding family from Virginia, he no doubt thought it natural to bring a group of "servants" to support his household in the distant Northwest. It is clear from these continued slave purchases by Fort Snelling officers that the army either regarded military posts as extralegal enclaves or chose to ignore territorial limits on slavery—or, perhaps, did not consider the question at all. Because there were few people available for hire as servants on the Upper Mississippi, military officers might turn to slavery as a solution, but some officers owned slaves before and after being stationed on the frontier. The region that would become Minnesota—the name was not even used until the 1840s—was part of Michigan Territory and, practically speaking, beyond the reach of the American judicial system in the 1820s. Judicial officials and functioning courts were hundreds of miles from the junction of the St. Peter's and Mississippi Rivers. Any responsibility regarding the legality of slavery in the region of Bdote lay with the army until the 1830s, and the army simply ignored the issue.[20]

Others arriving at Fort Snelling with slaves included surgeon John Emerson, who in 1834 brought along his male servant Etheldred—known as Dred—who later took the last name Scott. Dred married a woman named Harriet, one of Lawrence Taliaferro's enslaved servants, in 1837. Their marriage was approved of by both owners, and the Indian agent himself performed the ceremony. This would have been highly unusual in traditional slaveholding states.[21]

Enslaved people on a frontier post like Fort

Lawrence Taliaferro's list of the people he enslaved, with the notation, "21 freed from slavery 1830–40–43." *MNHS.*

Eliza
Frederick
William
Horace
Thomas
Samuel
Jerry
Homestead
John
Horace S—
Lizzie
Betsie   William E
Harriet   Turner
Susan   Turner
            Wyatt
            Phillis
            Lucy
            Charlotte

21 freed
from Slavery
1839–40–43

Dred and Harriet Robinson Scott, as illustrated in *Frank Leslie's Illustrated Newspaper,* 1857. *MNHS.*

James Thompson, born in Virginia in about 1799, arrived at Fort Snelling in 1827, brought by his enslaver, John Culbertson, who was the post sutler. Culbertson sold Thompson to Captain George Day, and Thompson served the officer until 1837. During that time, he interacted frequently with the Dakota, learning their language and becoming familiar with the territory surrounding the fort. While still in bondage, he married a daughter of Cloud Man (Maḣpiya Wiċaṡṭa), leader of the village at Bde Maka Ska (then Lake Calhoun), thus connecting him with whites like Lawrence Taliaferro, Seth Eastman, and Henry Sibley, all of whom had also married daughters of Cloud Man. In 1837, Methodist missionary Alfred Brunson, impressed with Thompson's linguistic abilities and overall character, raised funds among Ohio abolitionists to purchase Thompson's freedom. He hoped that the newly freed Thompson would help his own efforts to establish a Methodist mission at Kaposia. The mission failed, however, and Thompson seems to have fallen out with Brunson, who later accused him of being "unfaithful." He and his wife, who anglicized her name to "Mary," continued to live in the community on the east side of the Mississippi and later were among the early residents of St. Paul.[i]

Snelling worked inside, outside, and wherever they were needed. They did not work like southern field hands raising cash crops of rice or cotton; instead of facing the rigid controls of repressive slave laws and slave patrols in the South, they were surrounded by hundreds of miles of Indigenous territory, a barrier to escape for most. (But Joseph Godfrey, born probably around 1830 to the enslaved woman held by Bailly at Mendota, grew up in a household that spoke Dakota and French; he escaped to live with the Dakota in the late 1840s.) As valued workers, particularly for officers and officials with families, enslaved men and women would have been a recognized part of the fort's community, even if at the lowest level. Men performed heavy labor like chopping and hauling wood and tending gardens and livestock, while women cooked, cleaned, and cared for children (as well as doing any necessary chopping, hauling, and gardening), all work intended to allow their owners to live in a style conforming to their social status.[22]

The Scotts' marriage would have remained an interesting but obscure story had they not later brought suit in federal court in Missouri seeking their freedom. Their case was based largely on the ambiguous and contradictory status of people in bondage held in a federal installation in nominally free territory, and others had successfully argued similar cases in the Missouri courts. But in Dred Scott's case, which was heard as sectional conflict over slavery increased, though the suit was initially successful in the Missouri District Court, the lower court's decision was appealed to the US Supreme Court. In *Dred Scott v. Sandford*, issued in 1857, the Supreme Court ruled that the Scotts had no right to sue because they were not citizens—and that even freed slaves could not be citizens. If the status of enslaved people as property was based on who they were regardless of where they were taken or kept, then no state could outlaw slavery. This ruling had serious implications for the nation as a whole: it upset the delicate balance of compromises regarding slave and free states that had preserved the Union.

Some of the ugliest aspects of slavery also appeared at Fort Snelling. Lawrence Taliaferro mentioned in his journal that his servant girl, Eliza, gave birth to a daughter, Susan, in February 1831—with no indication of who the father might have been (it may have been Taliaferro himself) or what happened to the child. Another disturbing incident recorded by the agent involved the slave of his subagent James Langham. One morning, the Langhams' three-year-old daughter went missing and was later found unconscious, concealed under a pile of hay in the nearby stables. According to the agent, a "maid servant" belonging to the Langham family named Mariah eventually confessed, probably under duress, to injuring the girl but not meaning to kill her. Mariah was also implicated in several other incidents, including a house fire and the death of another of the family's children; no evidence survives, and she did not confess to these charges. Taliaferro had no doubt regarding Mariah's guilt, writing, "I was sure of this from the moment the little girl was missing that this vile negro was guilty." Mariah was soon held in chains: "an Iron collar, ball & chain & hand cuffs." The agent never mentioned her eventual fate, but as Langham traveled often to St. Louis, he may have sold her there. What motivated Mariah's actions, if she in fact did these things, we can only speculate, but the incident illustrates that relations between slaves and enslavers could be as fraught with violence on the frontier as in other regions. Years later, in about 1848, the body of an enslaved woman from Fort Snelling was found floating in Pig's Eye Slough, not far downriver from Fort Snelling. She had been brutally beaten to death and thrown into the river. The officer who owned her, not named in the sources, denied any knowledge of the event and was never prosecuted.[23]

To a large extent, Fort Snelling's military population mirrored that of the United States', carrying into the Upper Mississippi region many of the racial and social attitudes of the country as well as the institution of slavery. The ideas of race and white supremacy were accepted by much of this society as the natural order of things. These beliefs justified and directed both the nation's tolerance of slavery and the disinheritance of the Indigenous people of North America. Fort Snelling included the former, and it was a direct instrument in the latter. The practice of slavery would continue at Fort Snelling, in the free territory of Minnesota, into the 1850s.[24]

The very presence of the fort was causing change in the area of Bdote. A new community formed by whites, people of mixed ancestry, and Native people developed within the military reservation. Because the reserve was originally seen as a limited cession by the Dakota, allowing the army and Indian agency to occupy the area for the support of the post, military and political leaders had no intention of allowing it to become a focus for immigration and colonization of the region. The area was part of the Mdewakanton's ancestral lands, and they used the resources and visited sacred places as they always had; they treated the residents of the fort and the Indian agency as guests. The army was, in general, leery of allowing civilian settlements near frontier posts. The presence of white settlers, their acquisitive ways, and above all, their whiskey, were viewed by the army as ruinous to morale and harmful to relations with Native people. But every post commander, particularly in a place as isolated from direct oversight as Fort Snelling, was prone to use his discretion in such matters, and Josiah Snelling did so as well.[25]

In fact, events occurring well outside the immediate area led to the arrival of unexpected foreign refugees. In 1811, a Scottish nobleman, Lord Thomas Douglas, 5th Earl of Selkirk, determined to establish a colony on lands he had been granted around the junction of the Assiniboine and Red Rivers in what is today the Canadian province of Manitoba and the city of Winnipeg. The earl's purpose was basically twofold: the colony would serve as a refuge for displaced Scots and Irish tenant farmers, and the agricultural community that those farmers built would support the traders of the Hudson's Bay Company, in which Selkirk was a major shareholder. From the beginning, this scheme had problems. The location was extremely isolated and accessible only over long water routes, either through Canadian territory (via the York Factory on Hudson's Bay, the Nelson River, and Lake Winnipeg) or through US territory (via the Mississippi, St. Peter's, and Red Rivers). Both routes were closed from late autumn until late spring. The climate was harsh, far more extreme than anything the prospective colonists had ever faced. And finally, the fur traders of the North West Company were covertly and overtly opposed to its formation. They saw the presence of farmers as an existential threat to their trade. For these reasons and others, the Red River Colony began losing inhabitants within a few years of its founding, and the newly established American fort at the confluence of the St. Peter's and the Mississippi was an attractive immediate destination for those determined to leave.[26]

Selkirk had sent agents into Europe to recruit potential immigrants for his settlement, and they signed on just about anyone willing to go, including such nonagriculturalists as Swiss watchmakers. Among these were the families of Abraham Perret (or Perry, as he later anglicized his name) and Samuel Shadecker. Both had arrived in the colony around 1820 and soon regretted their decision to immigrate. "Though some of them were poor in their former homes," recalled Shadecker's daughter Ann, "they had at least comfortable dwellings and occupations which would give them bread. Here they had nothing to look forward to but destitution, trouble and toil."[27]

The Shadeckers were among about one hundred other Swiss and Scottish colonists who abandoned the Selkirk settlement in the spring of 1823. After a long, dangerous trek that included uncomfortable encounters with western bands of Dakota, they arrived at Fort Snelling in the autumn. They were among the first of what would be a regular flow of "Selkirkers" through the post. Most of them would move on as soon as they could obtain transportation to established towns like Prairie du Chien and Galena, but Colonel Snelling and Abigail evidently felt sympathy for those who wished to stay near the fort. The post was still under construction at this time, so perhaps the colonel also saw Samuel Shadecker as an able carpenter or mechanic to help with the project. Abigail was happy to take on thirteen-year-old Barbara Ann Shadecker, known as Ann, as a nanny for her rambunctious boys. The colonel allowed the rest of the family to live for several years in the barracks that Colonel Leavenworth had constructed at Coldwater Spring in the summer of 1820. They moved on to Galena in 1825, but Ann remained with the Snellings, eventually marrying an officer of the 5th Regiment, Joseph Adams.[28]

Severe winter snows and spring floods in 1826 caused great destruction in the area around Bdote and were equally devastating to the Red River Colony. Another large group of former colonists migrated through Fort Snelling in the summer of 1826, including the family of Abraham Perry. Again Colonel Snelling allowed the refugees to remain in the area around the fort. While most continued down the Mississippi, the Perry family began farming on the military reserve in the area of Camp Coldwater. Abraham Perry had managed to bring with him a small herd of cattle, which he gradually expanded over the years. His wife, Mary Ann, gained a reputation as an accomplished midwife. The Perry family also grew to include six daughters and one son. By 1837, the elder Perrys were regarded as patriarchs of the civilian population living on the reserve. As French-speaking residents, they would have fit in well with the fur-trading community already in the region, and the Perrys' middle-class European origins would have made them acceptable company for the official society at the fort.[29]

From the years 1821 through 1835, some 489 refugees from Selkirk's colony passed through Fort Snelling. Aside from foreign-born soldiers, they were among the first European immigrants intending to stay in the region, but they were by no means the only or even the dominant

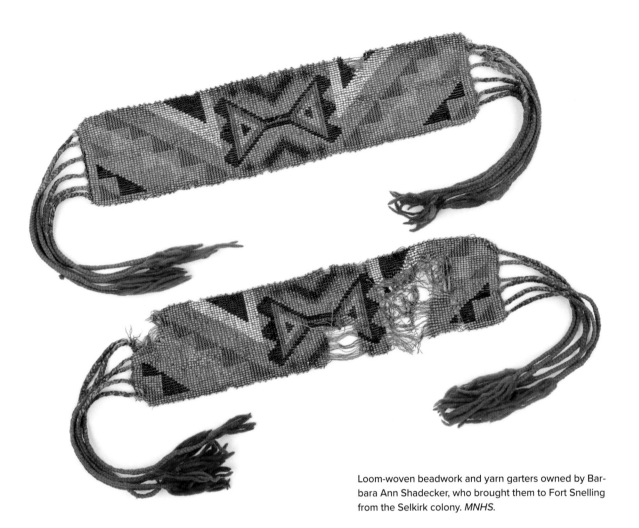

Loom-woven beadwork and yarn garters owned by Barbara Ann Shadecker, who brought them to Fort Snelling from the Selkirk colony. *MNHS.*

civilian population on the military reserve. The area around Coldwater Spring began to draw other civilian residents with deeper roots in the Upper Mississippi region. With fresh water, adequate wood, and nearby grazing for animals, it was an attractive place to stay. It is not clear how the Dakota, for whom the spring remained a sacred place, felt about this.[30]

Among the first to make a permanent home there was fur trader Benjamin Baker, who built a stone trading post and house near the spring in the 1820s. His wife was the daughter of the English trader John Stitt and an Ojibwe woman related to chiefs Hole-in-the-Day and Strong Ground. Others connected to the fur trade and the Indian

agency were also permitted by Colonel Snelling and his successors to establish themselves near the spring. These included Jacob Fahlstrom, probably the first Swede to reside in the region that became Minnesota. His wife was Marguerite Bonga of the African-Canadian-Ojibwe family long active in the fur trade. Marguerite was the sister of Rat's Liver or Rat's Heart, a prominent Ojibwe leader from the Mille Lacs area. Given the many connections to Ojibwe families, it is not surprising that members of that tribe often camped at Coldwater Spring when making their usual visits to Fort Snelling and the Indian agency there.[31]

The site also included the homes of two important employees of the agency: Antoine Pepin and

Joseph Rech or Roesch. A treaty signed at Prairie du Chien in 1825 promised both the Dakota and Ojibwe access to a blacksmith, to be located near the St. Peter's Agency, who could repair their metal tools and weapons, particularly traps, spears, and guns. Pepin was the agency smith for a time, and Roesch was his "striker," or assistant. Both also had connections to the fur trade and were married to women with mixed ancestry. Pepin's wife was Marie-Marguerite Hamelin, and Roesch was married to Suzanne Grant, the daughter of Peter Grant, a prominent North West Company trader, and Clear Sky (Ahdikacongab). Pepin worked at the Indian agency for nearly ten years when he was replaced by Oliver Cratte in February 1836. According to Taliaferro, Cratte was a skilled armorer, better able to repair guns, "just such a man as will suit the Indians who are delighted."[32]

—

All during the 1820s, European Americans, both military and civilian, recorded friendly interactions with the Mdewakanton Dakota. While these interactions are recorded by the whites, they suggest ways to imagine a Dakota perspective. Philander Prescott described how he met Spirit in the Moon, the daughter of Man that Flies, when she visited the sutler's store to trade quillwork or moccasins. "Her appearance and conduct attracted my attention and in fact the young woman got acquainted with all the officers, ladies of the fort and she became very much respected by all." After several years, Prescott married her, providing gifts to her family according to Dakota custom; he helped to support her family for many seasons, while they widened his trade network. They were also wed in a Christian service in 1837 and remained together for over forty years. Charlotte Ouisconsin Van Cleve, the daughter of Captain Clark, recalled that she and her brother frequently played with local Dakota children and knew many of the adults as well. They spoke

enough Dakota that, on one memorable and bitterly cold morning, as they impulsively pursued an injured wolf that had made off with their trap, they could enlist a passing Dakota boy to carry on the hunt while they returned to warm clothing, breakfast, and worried parents.[33]

Among the most influential relationships were those of Lawrence Taliaferro. Though the agent would eventually have a Dakota wife and child, his earliest influence came not through his own relations but those of his interpreters Scott and Duncan Campbell. Like so many men in the fur trade, the Campbell brothers were of mixed Native and European descent. Their father, John Archibald Campbell, was born in Londonderry, Ireland, around 1760 and immigrated to British Canada around 1790. He worked for a time in the British fur trade at Mackinac Island and later Prairie du Chien. Sometime in the 1790s, he married Ninese, a Wahpeton Dakota woman, and they had five children. After the region became part of the United States, he remained at Prairie du Chien and became a US citizen. In 1807, he was appointed Indian agent for the area, but the following year he was killed in a duel with a British fur trader. Archibald's eldest son, Duncan, grew up in Prairie du Chien working in his father's portaging business and the fur trade. He married a Mdewakanton Dakota woman about 1817. Duncan Campbell had extensive contacts in the Dakota community and among fur traders as well. He acted as an interpreter for Henry Leavenworth and was even given land by the Mdewakanton on which to live. Duncan's younger brother Scott Campbell also played a major role, acting as Indian Agent Taliaferro's primary interpreter. Scott was born around the year 1790, and in the early 1820s, he married Margaret Menagre, the daughter of a Menominee woman and French-Ojibwe trader Louis Managre.[34]

Though Taliaferro's understanding of Dakota language and society undoubtedly improved over the years, he was essentially ignorant of both when he arrived. The actions of Duncan and particularly his brother Scott as interpreters and ambassadors for Lawrence Taliaferro were vital to relations between the Dakota and the US government at Fort Snelling. Everything the agent told the Dakota was filtered through his interpreter, as were their statements to him. To a great degree, Taliaferro's success may have been due to Scott Campbell's ability to translate and smooth over the agent's statements. His diplomacy was as important as Taliaferro's.[35]

As early as 1820, the interpreters demonstrated their diplomatic ability. In August of that year, Taliaferro received a letter from General Henry Atkinson at Council Bluffs (in present-day Iowa) stating that a party of Sioux "who I have no doubt were Sioux of the White Rock on the St. Peter's crossed the Missouri and stole five public horses from this place on the 28th July, and on the 6th Instant some of the same party killed two of [fur trader Manuel Lisa's] People. . . . I have most earnestly to request that you will, as soon as practicable, investigate the matter, and cooperate with the commanding officer at St. Peter's to bring the murderers to justice and recover the horses taken here, and from the men that were killed." Given that the 5th Infantry had only a few hundred men, most of whom were busy building the new fort and few of whom were mounted, it was impossible for the army to comply with this request. The "Sioux of the White Rock" were also Sisseton, a band of Dakota that then had little contact with the Indian agency. Colonel Snelling, possibly on Duncan Campbell's advice, sent another Campbell brother, Colin Campbell, to the Sisseton asking for a council. Wishing good relations, the band sent emissaries and even left hostages at the fort to be exchanged for the men who had killed the traders. However, taking advantage of an inexperienced guard and the unfinished nature of the fort, the hostages escaped. Taliaferro

was away from the post at the time, and Colonel Snelling described the incident:

It mortifies me exceedingly that I am obliged to inform you of the escape of the two hostages by the carelessness and folly of the sentinel who was placed over them; the morning after you left the post they requested him to go with them outside the gate for a necessary occasion, he immediately passed them out without notifying the guard, or obtaining the permission of the officer of the day, as soon as they were out of the gate they started and ran, he fired upon them without effect, and the guard being unprepared for persuit [sic] they affected their escape.[36]

Philander Prescott witnessed the episode and later recalled that the sentry barely missed one of the escapees: "The ball passed so close to his head that he fell to the ground. The other, Maza Tonka, Big Iron, saw him fall and stopped and asked him if he was hurt, the other replied he thought not. 'Well,' said Maza Tonka, 'we can get away so up and be off for the sentinel is far behind.' So they started again and went clear."[37]

Following this fiasco, Colonel Snelling once again sent Colin Campbell to the Sissetons. Such was Campbell's power of persuasion—and perhaps the band's desire to establish good relations with the Americans—that they not only agreed to turn over one of the two men responsible for the killing of the fur traders, but they also sent a second man, an elder war chief and father of the other accused man, who volunteered to go in his son's place. Their arrival at the fort made a dramatic scene, as Colonel Snelling described it in a letter to Agent Taliaferro:

The unfortunate wretches were delivered up last evening with a great deal of ceremony and I assure you with affecting solemnity. The guards being first put under arms, they formed a procession in the road beyond the bake house, in front marched a sussitong [Sisseton] bearing a British flag, next came the murderer and the devoted chief, their

arms pinioned, and large splinters of wood thrust through them above the elbows, intended and understood to show us that they did not fear pain and were not afraid to die. The murderer wore a large British medal suspended to his neck, and both of the prisoners bore offerings of skins, etc. last came the chiefs of the Sussitongs, in the order they moved, the prisoners singing their death song and the Sussetongs joining in chorus until they arrived in front of the guard house where a fire being previously prepared, the British flag was burnt, and the medal worn by the murderer given up. The blacksmith then stepped forward and ironed him and he was conducted into the guard house. When the old chief offered his wrists to be ironed I told him it was not our custom to punish the innocent for the guilty, that he would be detained as a hostage and kept in confinement, but that he should be well treated and when the other murderer was taken he should be permitted to return to his band.[38]

Though Snelling immediately sent the supposed killer to St. Louis for trial, there were no witnesses or other evidence on which to try him. Secretary of War Calhoun felt that the surrender of the men at St. Peter's, along with the British flags and medals, represented sufficient recognition of American authority by the Sisseton Dakota, and he allowed the accused man to go free. Yet the Sisseton surrender of their men and British flags had little to do with fear of the small garrison at Bdote and more to do with the influence of Colin Campbell. Possibly as a result of this encounter, the Dakota of the upper St. Peter's and Red River regions were infrequent visitors at Fort Snelling, and they remained mostly outside the influence of the fort's Indian agent for several decades.

The 1820 incident was also an early demonstration of the conundrum at the heart of the fort's mission. The small number of poorly supported

troops made it impossible for the army to keep the peace between the tribes on the frontier or protect the Indian trade without the cooperation of the inhabitants of the region. In some ways, the American predicament resembled what the French had faced more than a century before. They did not have the numbers to impose their law and concept of justice in a vast, distant territory long governed by custom and kinship. Colonel Snelling was keenly aware of this situation. In 1824, he wrote Taliaferro, then on leave, regarding attacks upon traders located on the St. Peter's River "by a little band of marauders" led by "Markeemani."

> As I am restricted to defensive measures it is impossible to correct these abuses. To demand the surrender of the robbers, or the property, unsupported by a military force would only excite their laughter, and even supposing the robbers were surrendered, they could not be convicted by any evidence I could procure, or their own confessions out of court made in a language we do not understand. The truth is our judicial code is not adapted to these people, they should be dealt with as the Scotch say "by the strong hand," on the commission of the next murder, let me be authorized to proceed to their camp, inflict summary punishment on the offenders on the spot, and we should hear no more of it, this is the mode of proceeding best suited to their ideas and they would submit without a murder, but our present defensive attitude only invites aggression, if an Indian murders a White Man we are not to retaliate but appeal to the Courts of law, that is to say if the Indian will be pleased to appear before such a tribunal.[39]

The colonel's frustrations were obvious, as were his limitations for action. But he had other issues to address, as life within the walls brought continuing challenges.

Fort Snelling from the northwest, 1848. *Graphite drawing by Adolph Hoeffler, MNHS.*

5. Colonel Snelling's Fort

THROUGH THE YEARS OF JOSIAH SNELLING'S tenure at the confluence, the work of officers and soldiers remained difficult, and some of the trouble was of their own making. An understanding of the problems within the military, especially among the fort's officers, provides context for the larger developments in the region. And no matter what their squabbles might be, outside the fort's walls, the army still had to act as one.

In the early years of Colonel Snelling's command, morale was basically good. Both officers and enlisted men were focused on building Fort St. Anthony on the bluff above the junction of the rivers. Much credit is due Colonel Snelling, who guided the project with skill, cajoling the army for supplies and keeping his men fed, clothed, and mostly content while preserving good relations with the Dakota. But in all this he also relied on his officers.

With few exceptions, Snelling seems to have worked well with veteran "old army" officers like Major Josiah Vose and Captain Robert McCabe, whose backgrounds resembled his own. He had less regard for his younger subordinates, however, particularly those who were graduates of West

Point; he saw them as men of little practical experience. Indeed, the life of an officer on a frontier post like Fort Snelling, where gardening and wood gathering often supplanted soldiering, may have seemed disappointing and frustrating to a newly minted West Point graduate anxious to advance his career. As years went by and more officers came through the military academy, Snelling would have increasing conflicts.[1]

Indeed, relationships between officers could be fraught. A particularly nineteenth-century notion most evident among the officers at the post was the concept of honor. From the earliest days of the US Army during the Revolutionary War, officers saw themselves as "gentlemen," a status that required a high sense of personal honor. On the one hand, this meant maintaining decorum, good manners, and polite, chivalric conduct toward women, especially those regarded as "ladies." On the other hand, an action by a man of equal social standing that was regarded as insulting to the reputation of a gentleman or his family required a personal response. Demanding and receiving a formal apology might be sufficient, but in more extreme cases, individual

A percussion lock .54-caliber dueling pistol, converted from a flintlock. The pistol has a full walnut stock, an octagon-to-round barrel with a blade sight, engraved brass mountings, and an engraved, case-colored lock plate. It was made by "S. Homer," probably between 1800 and 1830. *MNHS.*

honor could only be "satisfied" through personal combat—a duel.

Though these concepts of honor and chivalry originated in aristocratic Europe, the practice flourished in the early years of the American republic. Since anyone might rise to the status of "gentleman," those who claimed this status could be very sensitive to any slight or disparagement. So prevalent was "the rage for dueling" in the Continental Army during the Revolutionary War that allied French officers, no strangers to dueling themselves, made note of it. The practice may have been exacerbated by what historian Ron Chernow describes as "a craving for rank and distinction that lurked beneath the egalitarian rhetoric of the American Revolution."[2]

These notions persisted well into the nineteenth century, though many inside the army and out found the practice appalling. Officially, the American military viewed dueling as a waste of valuable lives and a distraction from the army's role of preserving peace and defending the young republic. The army regulations of the era, adopted by Congress in 1806, specifically prohibited the practice.

Under Article 25 of the Articles of War, sending or accepting a challenge was forbidden. A commissioned officer could be cashiered—removed from the army—and a noncommissioned officer or soldier was subject to a court-martial and potential "corporal punishment," usually flogging. Despite these prohibitions, duels were not uncommon in the army in the era, and some officers even regarded the regulation as a "shelter for cowards." In the situation of a frontier post like Fort Snelling, a relatively small, isolated garrison virtually cut off from the outside world during the winter months, the opportunities for conflicts and their "gentlemanly" resolution were rife.[3]

The situation was further complicated for Colonel Snelling by the presence of his son William Joseph Snelling, who had moved to the post and joined his household in 1821. An ambitious and impulsive young man, "Jo" Snelling was the child of Josiah Snelling and his first wife, Elizabeth Bell, who died soon after the boy's birth on December 26, 1804. Though he was raised by his mother's relatives, Jo seems to have been a favorite of his father's and sought to follow Josiah's example by seeking a military career. He entered West Point at the tender age of fourteen but left in 1820, after just two years, possibly for disciplinary reasons. Jo headed to St. Louis, where he joined the Columbia Fur Company as a trader. He apparently found his way to the Red River Valley and spent the next winter with Wahnatah's band of Yankton Dakota; while there, he claimed to have become familiar with the Dakota language. He then accompanied Wahnatah the following spring to visit his father's new post at the confluence of the Mississippi and

St. Peter's Rivers. The details of this travel seem doubtful, as the journey from New York to the Red River would have taken several months. Also, Taliaferro's record of the arrival of Sisseton leaders at his agency in 1821 does not mention Joseph Snelling arriving with them—though he does appear on the agent's list of licensed traders at Traverse des Sioux on the Minnesota River in 1822.[4]

Young Snelling did, however, accompany the expedition of Major Stephen Long in 1823, along with the colorful Italian adventurer and self-styled explorer Giacomo Costantino Beltrami. Tasked with surveying the Red River Valley and the US border with Canada, the expedition stopped for a time at Fort St. Anthony to take on supplies, purchase horses, and hire interpreters. Beltrami, who was visiting the fort at the time, went along as supernumerary, a kind of tourist, really, while Jo Snelling joined the expedition as an "assistant interpreter" to the much more experienced Joseph Renville. A few weeks into the journey, however, Beltrami parted ways with Long. According to the Italian, it was not a happy separation: "[Joseph Snelling] also left the Major at the same time, not without violent altercation and went back to Fort St. Peter [Fort St. Anthony], by way of Lake Traverse. He quitted me in tears exclaiming, 'what will my father say?'" But if there was a conflict with young Snelling, Major Long chose not to acknowledge it. His report mentions that when the expedition took to boats on the Red River, he gave Joseph Snelling the assignment of returning some of the horses and soldiers to Fort St. Anthony.[5]

The years 1825 and 1826, however, saw a dramatic shift, both in morale at the fort that now bore Snelling's name and in Colonel Snelling himself. In May 1825, the colonel took an extended furlough. With his family, he traveled first to St. Louis and then to Detroit, visiting family members and engaging in private business, including the purchase of a farm near Detroit for $2,000 plus about $400 for repairs and improvements on the property. The Snellings returned to the Upper Mississippi rather late in the year, in early November. On their journey by keelboat from Prairie du Chien to Fort Snelling, unseasonably cold weather caught the family and their escort just upriver from Lake Pepin. Trapped by ice and snow in an early winter blizzard, they sent messengers overland to the fort, which sent a party of troops with supplies and transport to rescue them. They finally reached the fort on December 4, 1825. A few weeks later, Abigail Snelling, who had been in the final stages of pregnancy during this ordeal, gave birth to the couple's seventh child, Marion Isabella.[6]

The travel and narrow escape seem to have taken a toll on the colonel's health. He had long suffered from dysentery contracted during the War of 1812. This disease, usually caused by bacteria, was not uncommon in the army. It inflicted intermittent episodes of severe diarrhea and cramping on those affected and was usually treated with laudanum, a mixture of alcohol and opium. The treatment stopped the intestinal cramping but could, in turn, create other problems, such as severe constipation, which in turn was treated by enemas. It was also addictive. So life for Colonel Snelling was at times very unpleasant, if not painful, which may account for some of his behavior. It was also widely known at the time that the colonel, like many other officers, liked to drink, sometimes to excess. Ann Shadecker Adams, who had witnessed the colonel's actions while serving as the family's nanny, remembered, "Usually kind and pleasant, when one of his convivial spells occurred, he would act furious, sometimes getting up in the night and making a scene. He was severe in his treatment of the men who committed a like indiscretion. He would take them to his room, and compel them to strip, when he would flog them unmercifully. I have heard them beg him to spare them, 'for God's sake.'"[7]

The colonel's style of conduct cannot have

endeared him to the junior officers, some of whom were West Point graduates—with whom Jo Snelling socialized. One of these young men in particular would find himself in a head-on clash with Colonel Snelling. Second Lieutenant David H. Hunter graduated from West Point in 1822. The 5th Infantry was his first posting and very nearly his last. He likely arrived at Fort St. Anthony sometime in late spring or early summer of 1822; he first appears in regimental records as serving on a court-martial in June of that year.[8]

Hunter was, by most definitions of the time, a "gentleman," and during the first three years of his service with the 5th Infantry, he seems to have gotten on well with his colonel and the Snelling family. But sometime in 1826, young Snelling came into conflict with Lieutenant Hunter.

Though the cause of their dispute is not clear, fur trader Philander Prescott related in his memoirs, "During the last spring [probably 1826] a duel was fought between Lieut. Hunter and Col. Snelling's son, Joseph. The Col. heard that the parties had gone out to fight and sent a guard out and stopped them. The parties moved off and the guard went home and the combatants met in another place and exchanged shots. The Col.'s son had the end of one of his fore fingers shot off. This ended the strife and they went home."[9]

Later court testimony indicated that Joseph Snelling actually issued the challenge to Hunter, probably in May or June, and that far from discouraging the event, Colonel Snelling offered to

Second Lieutenant David Hunter later became Major General David Hunter, photographed by Matthew Brady during the Civil War. *Library of Congress.*

"take a shot himself" if his son should fall. Whatever the details, the event surely soured Hunter's relationship with the colonel and his family. Perhaps more telling is the fact that the post commander could have charged Hunter under Article 25, but chose not to.[10]

This duel was not an isolated event. Several months earlier, in his journal for February 6, 1826, Agent Taliaferro had noted, "An affair of honor settled near Fort Snelling this morning between two officers of the 5th Inf. Stationed at this post." And Prescott also remembered another duel that summer, fought between Lieutenant Baxley and Captain Leonard, the sutler: "They fired some 6 or 8 shots apiece. Lieut. Baxley got one shot, Capt. Leonard's second forgot to cock his pistol but Lieut Baxley missed. Seconds interfered and tried to reconcile matters but nothing would or could be done satisfactorily and at it they went again. I believe the 8th shot Lieut. Baxley's ball struck Capt. Leonard in the corner of the eye near the temple but did not kill him although it came very near killing him. Lieut. Baxley had 32 ball holes through his clothes but did not draw blood once."[11]

These duels in 1826 were evidently widely known at the post, and far from prosecuting the participants or discouraging the practice, Colonel Snelling seems to have encouraged it. Though a number of the participants were wounded, there is no record of any deaths as a result of the duels.[12]

Why this sudden outbreak of personal conflict in the garrison? The winter of 1825–26 was particularly harsh. Lawrence Taliaferro's journal reports temperatures in late January ranging from sixteen to twenty-four below zero; on February 14, the mercury refused to budge out of the bulb "graduated to -30 des. If this be not cold weather, what is?" Taliaferro also recorded a number of heavy, violent snowstorms, noting that "many persons have been frosted—Feet and Hands severely" and a number of cattle and horses killed. Work, travel, or any of the outdoor activities that had amused members of the garrison in more pleasant times would have been impossible. With little to do beyond the standard military routine, opportunities for quarrels and conflict were inevitable. An order issued in January 1826 hints at the state of affairs. It established a special guard "to report immediately to the Commanding Officer of the guard, all fighting or riot, all cases of candles being lit after hours, every instance of filth being thrown on the parade or men urinating in front of or against the barracks, in any instance of this kind the offender will be immediately confined.

No curtains will be allowed in barracks rooms, except where women are quartered."[13]

While some officers were taking out their animosity with pistols, another was enduring a legal proceeding. The court-martial of Lieutenant Phineas Andrews began in February 1826 and went on for months—through the cold, snowy winter, devastating spring floods, and a busy summer—finally concluding in August. Andrews had begun military life as an enlisted man, served in two wars, and received a battlefield commission during the War of 1812. He was charged with conduct unbecoming to an officer, including drinking and card playing with enlisted men, drunkenness on duty, and general incompetence. Colonel Snelling presided; most of the testimony seemed divided between enlisted men, who generally supported Andrews, and officers, including the post surgeon, who did not. Noting in his defense that he had risen through the ranks, Andrews concluded, "It has been my misfortune gentlemen, for it certainly is one, to have excited the displeasure and aroused the vindictive passions of some of my brother officers." The court found him guilty, recommending that he be cashiered.[14]

Lieutenant Andrews may well have been a hopeless drunk, but he was far from the only officer at the post with a serious drinking habit. Alcohol consumption was a problem in the army and American society in general, and Fort Snelling was far from immune. The trial also seems to have caused disputes among the officers and between junior officers and Colonel Snelling.

Lieutenant Andrews's trial ended on August 8, but in the week before its conclusion, several dramatic clashes occurred between Colonel Snelling and two of his junior officers, Second Lieutenants David Hunter and Nathaniel Harris. Hunter had been placed in temporary command of Company F to replace its lieutenant, who was absent "accompanying the Indian Agent into the Indian Country." (It seems curious that the colonel

July 31, 1826

Sir,

You taking advantage of your official situation to injure "F" Company and thru its present commander makes it my duty to call on you for the satisfaction of personal combat. I now therefore challenge you to meet me as soon as convenient. The law on this subject by your own construction, being only intended as a shelter for cowards.

You having publicly offered to give the kind of redress I wish to our Grey Headed sutler, when you had or supposed you had injured him and you having on another occasion, offered to meet any officer who might be offended at a certain order you threatened to give concerning dogs, all this induces me to believe you will be courteous enough by complying with my call as quick as possible.

I have the honor to be
With due respect,
Your obedient Servant,
David Hunter.

should have named Hunter to this position, given their personal animosity, but seniority and availability may have determined the choice.) But Colonel Snelling reached around his new acting commander, breaking military regulations, to assign two men from Company F as additional waiters for Lieutenant Baxley, who had just been made post adjutant. Hunter took this as a direct, personal insult. Given the history between the Snellings and Lieutenant Hunter, it was probably no surprise to the colonel when he received a letter from the lieutenant challenging him to a duel.[15]

How shocked the young lieutenant must have been when the colonel immediately placed him under arrest for violating Article 25. Had Colonel Snelling, knowing Hunter's temperament and past history of dueling, set him up for arrest, court-martial, and ultimately dismissal from the army? Hunter would have to await his trial to raise the question.[16]

Such was the colonel's ire that he placed Hunter under close arrest and eventually shipped him off for several months to Fort Armstrong, at modern day Rock Island, Illinois. The distance from Fort Snelling to the military post at Benton Barracks near St. Louis, where his trial would be held, and various emergencies during 1827, would delay Hunter's court-martial for over a year, an unpleasant time for the young man. But it allowed him plenty of leisure to consider his defense.

Meanwhile, further events centering on the Andrews case roiled the waters at Fort Snelling. On August 5, 1826, while the court was still deliberating, a heated confrontation took place in the officers' quarters. That afternoon, Lieutenant Hunter, though under arrest, called Jo Snelling into his quarters, and, along with Lieutenant Nathaniel Harris, they began to discuss the Andrews trial. In the course of the conversation, both Harris and Hunter implied that the charges against Lieutenant Andrews were untrue. Young Snelling took this as a deliberate insult against his father, who was president of the court. The colonel quickly learned of this incident and ordered acting adjutant Lieutenant Baxley to summon Lieutenant Harris to Baxley's quarters. There the elder Snelling accused Harris of having invited Joseph into Hunter's quarters to deliberately insult him. Harris denied this, and the conversation soon grew very heated. The enraged Colonel Snelling began to insult the young lieutenant in a very threatening way. Harris later testified that at one point he asked the colonel whether he, Harris, should regard the colonel's insults as coming from "Colonel Snelling or his [Harris's] Commanding Officer." At this point the colonel ordered his arrest—and then declared he "wasn't worth arresting."[17]

These conflicts between junior officers and Colonel Snelling drew the notice of army inspectors reporting to General Edmund P. Gaines. A report from August 1826 by Colonel George Croghan noted, "That harmony so desirable at every post and especially so at a frontier one, seems not to prevail here. The officer in command and some of his junior officers are at variance. He gives his orders, they obey them, tho' not without some grumbling and questioning of the correctness."

According to the report, much of the problem had to do with the general situation of the fort. Many of the younger officers felt their abilities were being wasted and their military skills deteriorating as the colonel pressured them to act as "builders and farmers."[18]

Croghan's overall comments regarding the frontier posts pretty well sum up Fort Snelling's situation—one in which the needs of supply overwhelm the requirements of soldiering: "What has been gained by this anti-military course to compensate the great loss of moral strength which has been sustained? Nothing, so far as I have been enabled to ascertain, that is of true value to the soldier. A few dollars have been added to the administrative fund, but at what a cost! Look at Fort Atkinson and you will see barn yards that would not disgrace a Pennsylvania farmer, herds of cattle that would do credit to a Potomac grazier, yet where is the gain in this, wither to the soldier or the government?"[19]

Despite Colonel Croghan's criticism, the variable climate of the region compelled these frontier posts to grow their own supplies—or face starvation. Even in good times, commissary supplies were irregular, and the year 1826 was particularly uncertain. The harsh winter was followed by dramatic and catastrophic flooding. Lawrence Taliaferro noted on April 23, "The high water and ice together has taken off all the houses on the St. Peter's and Mississippi near this post," and Dakota villages "20 miles up the St. Peter's" had been forced to seek higher ground. The lost buildings included "Suds Row," the buildings at the landing below the fort that housed the laundresses and their families. To make matters worse, while Colonel Snelling and his family were away on leave, a combination of bad luck and poor decisions by the acting commander at Fort Snelling, Lieutenant Colonel Morgan, had left the post short of supplies. In a report written to Quartermaster General Jesup in April, Colonel Snelling laid out his tale of woe (see sidebar).[20]

Conditions at the confluence deteriorated during Snelling's leave. On April 1, 1826, he wrote a letter to Quartermaster General Thomas Jesup explaining why—and displaying his bitter frustration.

When I left here on the 17th of May last, one hundred and fifty acres of corn had been planted, and about the same quantity of wheat sown; I had also fitted a wood boat, and detailed a party which brought to the landing about one hundred and fifty cords per week. When Lt. Col. Morgan assumed the command, being under the impression that drill and discipline were paramount to all other objects, he dismissed the party, and laid up the boat. The blackbirds which have always been troublesome enough, by far exceeded in numbers anything known in former years. . . . [T]hey descended in myriads, and when they alighted in a few minutes not a kernel of corn remained. . . . The wheat also was blighted, and after being stacked was condemned as not worth threshing. The haymaking was confided to Lt. Russel, who although an excellent man, and a most valuable officer, was the last person at the post I should have selected for that service, as he was bred in a city and knows nothing of the business. . . . Towards the close of the season Lt. Col. Morgan became apprehensive that the forage was insufficient, and purchased five hundred bushels of corn at Prairie du Chien. In this too he was unfortunate for by carrying sail in the night, the boat on board which it was freighted, filled and sank, occasioning a total loss of vessel and cargo. . . . The extreme length of the past winter and the great depth of the snow, has occasioned the loss of many valuable cattle. In addition to those you will find on my return, the Commissary has lost forty four and sixty have died belonging to individuals.[ii]

Such were the shortages that Colonel Snelling felt compelled to get a head start on the next season's crops: "To procure a sufficient supply of vegetables for the ensuing winter immediately, All parades and Military duties, guard mounting accepted [excepted] will be suspended and every man not on detail will be employed in gardening under the direction of their company officers."[21]

So much for drill and discipline and "soldiering."

———

But against this backdrop of infighting and bad behavior among some of the fort's officers between 1823 and 1826, the larger work of the command at the confluence had been playing out.

One of Lawrence Taliaferro's chief frustrations was his inability to stop the intermittent fighting between the Ojibwe and the Dakota. Bringing peace to the region was among the primary missions of his agency, and he could not understand why fighting continued despite promises and stated desires for peace by all concerned. He mostly blamed the traders for stirring up old animosities and saw their growing use of whiskey as a destabilizing influence on the tribes.

He was not totally wrong about this, but he also failed to account for the larger picture: the effects of American policy and the nature of leadership among Dakota and Ojibwe people. They were aware of what had happened among tribes to the east of the Mississippi. The US government's expansion into the Ohio Valley after the War of 1812 was followed by white encroachment into the Northwest (now Michigan and Wisconsin) in the 1820s. With this invasion came pressure by mining and timber interests on the lands of nearby tribes like the Sauk and Meskwaki in Illinois as well as the Ho-Chunk and Ojibwe in Wisconsin. These developments created unease among the Dakota and Ojibwe people of the St. Peter's Agency, particularly as disputes arose with

tribes to the east who were increasingly hunting on Dakota lands along the Des Moines River.[22]

Further, Taliaferro did not immediately comprehend that Dakota and Ojibwe chiefs lead by persuasion and example, not command. Any decision required consensus, and though a leader might agree with the agent that fighting should end, he could not guarantee that all those following him would accept the idea, particularly individuals with personal grievances or young men anxious to prove themselves as warriors. And though as the agent Taliaferro might use personal relationships and promises of gifts to influence the nations of the St. Peter's region, the troops at Fort Snelling did not have the manpower or means to enforce the peace, even on the military reserve itself. The agent's influence with the western bands, the Yankton and Sisseton, was even smaller. They continued what they regarded as defense of their hunting grounds against the Ojibwe, from the north, as well as the Sauk and Meskwaki, from the south. Ojibwe leaders, usually accompanied by members of their respective bands, were also frequent visitors to the agency, where they too received gifts and assurances from Taliaferro. The agent had a particular regard for Flat Mouth (Eshkibagikoonzhe), leader of the Pillager band: "he is a great stickler for etiquette and is absolute among his band." If not "absolute," Flat Mouth was certainly influential, bringing two hundred warriors to his initial meeting and negotiation with the Dakota at the St. Peter's Agency in 1821. In 1823, Taliaferro persuaded a number of Dakota leaders to make a visit to Washington and other eastern cities. The agent hoped the chiefs would help him persuade Secretary of War Calhoun to call for a grand council in order to ensure peace among the tribes.[23]

Agent Taliaferro and Colonel Snelling considered the trip a success. Calhoun agreed to the idea of a grand council, and, as Taliaferro had hoped, the trip impressed upon the travelers the numbers and power of the Americans. Colonel Snelling reported to Calhoun a year later, "The effect produced by the visit of their chiefs to Washington is wonderful. . . . [T]he power, wealth and numbers of the American people have been their constant themes, many of their stories approach so near the marvelous as to be discredited, such for example is the account of casting a cannon which they witnessed, and the magnitude of our ships. Old Black Dog shakes his head and says 'all travelers are liars.'" The main disappointment of the excursion was the fact that few representatives from western bands joined the group, and the journey was marred by the death of Chief Cloud of the Wahpekute.[24]

William Clark, the head of the Office of Indian Affairs, and Lewis Cass, governor of Michigan Territory (which then included what is now Wisconsin and Minnesota east of the Mississippi River), called a council of virtually all the tribes of the Upper Mississippi region to convene at Prairie du Chien in August 1825. Many leaders of both the Ojibwe and Dakota attended. In his opening remarks to the gathering, Clark assured the representatives of the tribes that the purpose of the council was not land cession, but guaranteeing peace among them by setting fixed boundaries. Such rigidly delineated borders, as used by Euro-Americans, were not part of the worldview of most Native American communities. Some of the chiefs involved in the negotiation feared that setting such boundaries would actually create more problems. Coramonee, a leader of the Ho-Chunk, said that his people used lands in common with many of their neighbors, and "It would be difficult to divide it. It belongs as much to one as to another." Noodin (the Wind), an Ojibwe leader from the region between the St. Croix River and Mille Lacs Lake, feared that "in running marks round our country or in giving it to our enemies it may make new disturbances and breed new wars." Despite these misgivings, they eventually agreed to accept boundaries. Negotiating the border locations proved difficult, however. In many cases their territories overlapped or were otherwise uncertain. The boundary eventually agreed upon between the Ojibwe and Dakota included landmarks familiar to both tribes, locations like "the place they buried the eagles" or the "cedar the Sioux split." The Ojibwe representatives, probably more experienced in dealing with white officials, insisted on an article that provided flexible boundaries in areas where hunting rights were either disputed or traditionally shared. Article 13 stated, "The Chiefs of all the tribes have expressed a determination cheerfully to allow a reciprocal right of hunting on the lands of one another, permission being first asked and obtained."[25]

Though no lands were given up by this Treaty of Prairie du Chien of 1825, the setting of boundaries that clearly identified specific areas with specific peoples essentially started the process. It laid the groundwork for future treaties in which tribes could be coerced into giving up some or all of their lands. Many of these later treaties would refer specifically to the lines drawn at Prairie du Chien in 1825.[26]

Taliaferro had high hopes for this treaty, most of them quickly dashed. A mysterious fever struck many of the most prominent chiefs among both the Dakota and Ojibwe while they were returning from the council. William Laidlaw, the Columbia Fur Company trader "who was employed to take the sick Indians of the Yanktons, Wahpeton, Sissetons and Wahpekute to Lake Traverse," reported weeks later that many were still ill and that Yankton chief Wahnatah had died. "The Indians stated that they had been poisoned at Prairie du Chien," Taliaferro noted, "by drinking mixed whiskey and sugar and what confirms them [in] this belief is that all who drank of it have been taken sick and many have died. They attach great blame to Laidlaw and [Joseph] Renville for inducing them to attend the Council." Though Taliaferro tried to limit the damage by distributing

Encampment at Prairie du Chien for the Treaty of 1825, which was signed in September. *Lithograph by Lehman & Duval Lithographers, LOC, after a painting by James Otto Lewis.*

small gifts of whiskey and goods to those who had lost relatives or been taken ill, the incident harmed his already tenuous relations with Dakota and Ojibwe people living beyond the immediate area of Fort Snelling.[27]

Another problem was the boundary set by the treaty. Before, Fort Snelling had been seen as lying between the Dakota and a loosely defined, disputed zone between the Dakota and Ojibwe running from the Mississippi to Mille Lacs. The new treaty, however, put the fort and the agency clearly in Dakota Territory—while the agency was still responsible for the Ojibwe bands of the St. Croix and Mille Lacs and the Upper Mississippi regions. Taliaferro

noted that large numbers of Ojibwe, as many as "2300 or upwards, attach themselves permanently to the St. Peter's Agency." Like the Dakota, they viewed the agent as their intermediary with the US government and the traders. Groups from various bands of Ojibwe visited the fort and agency frequently, especially in the summer. At times, many Ojibwe would be present at the agency, and tensions with the Dakota were constant. On January 6, 1826, Bears Heart (Makode'), an Ojibwe leader, visited the agency, and the following day the Dakota war chief of Black Dog Village, Cloud Man (Maḣpiya Wiċaṡta, who would soon become Taliaferro's father-in-law), visited Taliaferro with the complaint that the Ojibwe "had not kept their word for they had crossed the line and had hunted all the small and large game off and had not left any to enable him to pay their credits to the traders."[28]

The agent continued to note to his superiors that delegations from both tribes were visiting the St. Peter's Agency, but there were insufficient stores to satisfy them all: "Since the Treaty at Prairie du Chien it is certain from the report of this deputation of the Chippeways that all their nation on the waters of the Mississippi above Prairie du Chien—will visit this Agency in future—say between 700 and 800 men will annually visit me. . . . The appropriation is now too small by one third and will require an addition of one half to meet expenses of this addition." Far from preventing war between the tribes, the treaty seemed to have created a situation that would guarantee it.[29]

Perhaps to relieve this situation—or, as Taliaferro supposed, to placate the American Fur Company—the Office of Indian Affairs decided to remove the Ojibwe of the Upper Mississippi, the St. Croix, and the Mille Lacs areas from the St. Peter's Agency and place them under the Michigan superintendency of Henry Schoolcraft at Sault Ste. Marie, about six hundred miles by river and portage to the east of Fort Snelling. While Schoolcraft might decide to meet these bands at La Pointe on Lake Superior, the Ojibwe would still have to travel a considerable distance to receive annuities, lodge complaints, and carry out other diplomacy.

This move was immediately unpopular with the Ojibwe. Taliaferro recorded the protests of Little Six (Shagobe) of the Rum River band, who said that when his people were trading with Great Britain, "We could not see the light of their fire, it was so far off. We got no tobacco and had to smoke leaves. My father, when we heard of your fire, we soon saw your light and came here to see you. . . . You tell us that Mr. Schoolcraft and Mr. Johnson will be good friends to us, this may be but

they are too far off—we have too many portages to make and we have to remain too long from our families and get nothing when we go." Flat Mouth, the influential leader of the Pillager band, was even more outspoken: "The news which you have told me this day has hurt my feelings very much. I hardly know which way to look. . . . I can tell the truth as well as a white man and when I speak all believe me. My father, we made the road to your house that we might forever bury our war club and remain at peace with the Sioux, and now our road is stopped. I shall no longer be able to keep the peace and the war will commence again, worse than before and the plains will be covered with blood."[30]

Ojibwe leaders clearly felt that their people's interests were harmed by this seemingly arbitrary and pointless move, but their trust would be further strained by dramatic events just days after Flat Mouth's speech.

Around the middle of May 1827, a large number of Ojibwe, including women and children, arrived at Fort Snelling. They camped at Coldwater Spring and along the river below the fort. As often happened on these occasions, Dakota and Ojibwe people danced, feasted, and played games together. Because the Ojibwe were no longer getting goods from the agency, they exchanged birchbark canoes and maple sugar for trade goods with the Dakota and residents of the fort. All seemed well until the evening of May 27, when suddenly and inexplicably, a violent, deadly incident occurred.[31]

A group of Dakota warriors had been visiting the Ojibwe camped below the fort. As these men were leaving, they turned abruptly and fired their guns into the tents of their hosts. Two Ojibwe were killed outright and a number of others wounded, some fatally. What might have motivated such an attack was and remains unknown. The injured were brought to the Council House, where Taliaferro described "such a scene I have not witnessed before or to the same extent." Both the agent and Colonel Snelling were outraged at what seemed an unprovoked, cowardly assault, and they were mortified that it was carried out practically under the walls of the fort, upon a people supposedly under the protection of the United States. According to Snelling's account, Strong Earth, a chief of the Sandy Lake Ojibwe, berated both men in the Council House.

> You know that two summers ago, we attended a Great Council at Prairie du Chien, where by the advice of our white friends, we made a peace with the Sioux. We were then told that the Americans would guarantee our safety under your flag. We came here under that assurance; But father look at your floor, it is stained with the blood of my people, shed under your walls. I look up and see your flag over us. If you are a great and powerful people why do you not protect us? If not, of what use are all these soldiers?[32]

The colonel probably shared Strong Earth's opinion. In fact, he had already given orders to find those responsible. On the previous day, he had sent one hundred troops under Major Fowle, with William Langham and Scott Campbell acting as interpreters, to confront Shakopee's village: "Major Fowle will on arrival, demand the delivery of the men engaged in the murder of the Chippewa last evening under the walls of the garrison, if refused he will resort to force."[33]

The show of force, and likely also the persuasion of Scott Campbell, had immediate results. Four men, two from Shakopee's village and two others from Little Rapids, were given up to the soldiers. Rather than trying the accused at Fort Snelling or sending them off to St. Louis for trial, Colonel Snelling resolved that the situation demanded immediate retribution. He determined that the Dakota men should be given to the Ojibwe for punishment. This decision was somewhat out of keeping with what had happened previously, and Philander Prescott noted that there

was no certainty that the men handed over were the actual killers. On May 29, according to Prescott and Anna Adams, the prisoners were taken to the prairie just outside the fort, in view of the garrison and before a considerable number of Dakota. Then, one by one, the Ojibwe gave them a chance to run for their lives, coolly shooting them down. Afterward, the Ojibwe mutilated the bodies of the Dakota men, as if killed in battle, and threw the corpses into the river.[34]

The Indian agent's account of the executions was terse and somewhat detached. Perhaps Taliaferro could not bring himself to witness the killings. A few days later, most of the Ojibwe departed the area, with an army escort, while some of the more seriously wounded remained at the fort, accompanied by a few relatives. A number of them died over the following weeks. Taliaferro, meanwhile, sought to smooth things over with the Dakota and believed he had succeeded. But at least some of the relatives of the executed men held Taliaferro responsible for their killing and the manner of their deaths.[35]

It is not clear what Scott Campbell told the Dakota—whether they assumed the men turned over to the soldiers would be tried by whites or held as hostages. But now they had witnessed the execution of their relatives by their longtime enemies. While some "friends of the Sioux" questioned Snelling's decision to give the men over to the Ojibwe, others like Philander Prescott, Superintendent Clark, and General Atkinson supported Snelling's actions. Whatever their opinion, however, no one could still believe that the 1825 treaty had succeeded in bringing peace between the Ojibwe and Dakota, and events over the next few months would reflect growing unhappiness among all the tribes of the Upper Mississippi Valley.[36]

Two developments at Fort Crawford began the problems. In July 1826, soldiers there had taken into custody two Ho-Chunk men accused of killing the family of a French Canadian man north of

Prairie du Chien, then transferred them to Fort Snelling. And in August 1826, the army decided to abandon Fort Crawford. The post was located on low ground, susceptible to flooding and in poor repair. Further, in the aftermath of the 1825 treaty, an army presence was seen as unnecessary. The two companies stationed there were reassigned to Fort Snelling in October 1826, leaving only a few men to care for the guns and ammunition until those could be transported upriver.[37]

By the summer of 1827, however, the Ho-Chunk and the Sauk and Meskwaki were feeling increasing pressure from Americans, particularly lead miners and illicit timber operations on tribal lands. In July, a small band of Ho-Chunk led by Chief Red Bird attacked a white family near Prairie du Chien. They may have been responding to the false rumors—possibly a genuine misunderstanding, but possibly spread by relatives of the Dakota men killed in May—that the Ho-Chunk men still imprisoned at Fort Snelling had been executed; in addition, there were reports of attacks on Ho-Chunk women by river crews. A few days later, two keelboats traveling downriver from Fort Snelling were fired upon about fifty miles upriver from Prairie du Chien. The attack took place at night, and there was a long and intense exchange of gunfire between attackers on shore and the soldiers and boatmen on the river. Two of the boatmen were killed and four wounded, and though they claimed to have inflicted heavy losses on the attackers, there was no way of knowing how many of them the whites may have hit. More alarming for the garrison at Fort Snelling was not knowing exactly who had been attacking the boats. Were they Ho-Chunk or Dakota—or, worse still, both?[38]

As with the incident in May, Colonel Snelling reacted decisively to this perceived threat to the post's supply line. Four companies, about half of his command, were immediately ordered onto the remaining keelboats, and with the colonel himself in charge, they embarked on July 9 for

Prairie du Chien, where the government was soon assembling a force of hundreds of regulars and militia from Missouri, Illinois, and other parts of Wisconsin to fight the "outbreak." The threat from the Ho-Chunk ended when Red Bird and other leaders surrendered, but the great fear for Colonel Snelling—as well as Agent Taliaferro and other whites throughout the region—was that the Dakota and Ojibwe would join the Ho-Chunk, creating a more general war on the Upper Mississippi that the army had few resources to fight.[39]

At this time, relations with the Mdewakanton near Fort Snelling seemed mixed. On June 9, Taliaferro reported that Black Dog had sent out seventy of his young men to find a white child who had wandered away from the fort. While Taliaferro and others from the post had been looking for the little boy for more than a day, the Dakota found him within an hour and a half, "So my Indians are exempt from all blame by his absence," the agent wrote. "The unpardonable neglect of the Mother caused all this anxiety." But soon after news of the Ho-Chunk attack arrived, Taliaferro became aware that relatives of the men executed in May were seeking retribution. It was reported to the agent that an uncle of one of the men who was shot by the Ojibwe had held a feast for relatives of the dead man at Shakopee's village and promised all those attending that he would kill Taliaferro. Whether merited or not, the atmosphere of alarm among those remaining at the garrison and civilians living in the area of the fort became so strong that, as Taliaferro recorded it, "The Families generally near the fort, left their dwellings this evening and went into the Fort for security believing that an attack might this night be expected by the Indians." And though he and the subagent James Langham remained at the agency, Taliaferro was concerned for his own safety, having been warned "to be on my Guard and not to go far beyond my house . . . as I had been threatened, I was best to be on my look out." There was also concern that Wabasha's band, who had close ties to the Ho-Chunk, might take part in the fighting.[40]

But none of these fears proved justified. Despite the unhappiness among some bands and the ongoing tensions with the Ojibwe, the majority of Dakota maintained relationships with the traders and the Indian agent. Many of their leaders had visited the American capital in 1823 and had a clear idea of the size of the United States and the unlikelihood of a successful war against it. Though circumstances were not ideal, the Dakota could continue to live in much the way they had in the past, despite the fort and its community.

In early August, Colonel Snelling and the four companies sent to Prairie du Chien returned to Fort Snelling. The expedition was perhaps the most intense military activity in the 5th Regiment's time on the St. Peter's, but it served as only a short distraction to the ongoing turmoil among the officers of the regiment. Lieutenant Hunter's case was still unresolved, and another dispute with the colonel seems to have erupted, this time with Lieutenant Baxley. In an extraordinary letter to an unknown officer, copied in the colonel's journal, he accepted a challenge from Baxley.

St. Peters Sept. 10, 1827

Sir:

I have contrary to my duty and my principles and to gratify the bad passions of a bad man, consented to wave my rank to Lt Baxley and expect to receive a message from him today. As you have agreed to appear on the field as my friend, I think proper to dictate to you the following terms of combat, from which I will not depart.

The duel shall be fought at four paces with pistols, and the firing shall continue until one of the parties is killed or disabled. I do not go out for a show, and I will have no spectators, or surgeon. I will consent to no reconciliation, or shaking of hands. When I think a man a rascal I never take his hand.

J. Snelling[41]

On October 1, 1827, the steamboat *Josephine* arrived at Fort Snelling carrying General Gaines, commander of the Western Department. He was on a quick inspection tour along with the army paymaster. Previous reports regarding morale at the fort and issues with Colonel Snelling's commissary returns may have prompted this visit, as well as a determination to resolve the case of Lieutenant Hunter. The Snelling family evidently anticipated a departure from the post, as they had begun disposing of personal items in August, and family tradition relates that the colonel was being considered for appointment as governor of Florida.[42]

The following day the *Josephine* departed for St. Louis with the general, the Snelling family, Lieutenant Hunter, and officers ordered to attend the latter's court-marital. The steamer made record time, arriving at St. Louis on October 9.

General Gaines had issued orders for the court-martial on October 2, dating them from Fort Snelling and naming Colonel Henry Leavenworth president of the court. The remainder of the court, essentially the jury, was made up of officers from the 1st and 6th Infantry Regiments stationed at Jefferson Barracks, the site of the trial.

Proceedings began on October 16. Colonel Snelling, having brought the charges against Lieutenant Hunter, acted as prosecutor—"judge advocate"—and as a witness for the prosecution. Despite the presence of his old rival Colonel Leavenworth as president of the court, Snelling began with confidence, laying out the case as a cut-and-dried violation of Article 25. He brought forth witnesses to testify to the authenticity of Hunter's letter challenging the colonel and confirming Snelling's explanation of events before and after the letter.

Acting in his own defense, David Hunter used his cross-examinations to shape a different version of events, emphasizing Colonel Snelling's repeated statements regarding personal combat.

July 1st 1823. Prideville farm near
St Louis Cost — — — — $2000.00
" " " Recording deed 2.50.
" " " Taxes for 1823' 4. 10.89.
Octr 1st 1825. Expence of hauling corn
" " to St Louis 5.25.
" " " Taxes for 1826. 3.06.
$ 2021.70
Cr By rent corn 1825 $25.
Do in 1826 $25 — 50.00
1971.70

Sir St Peters Septr 10th 1829
I have contrary to my duty & my principles
and to gratify the bad passions of a bad man,
consented to waive my rank to Lt Baxley & expect
to receive a message from him to day. As you have
agreed to appear on the field as my friend, I think
proper to dictate to you the following terms of
combat, from which I will not depart.
The duel shall be fought at four paces
with pistols, & the firing shall continue until
one of the parties is killed or disabled.

I do not go out for a show, & I will have
no spectators, or surgeon. I will consent
to no reconciliation, or shaking of hands.
When I think a man a rascal I never
take his hand. Snelling
To Lt Green
Octr 11th 1827. Took lodgings at Mrs
McKnies for Mrs $4 pr week. James $2.
Marion $2. Mary (Slave) $2. Total. $10.00
Wood & candles to be furnished by me

May 29th 1825.
Purchased a ½ Section of land
in Illinois of Lt Green for — $20.00
Taxes to the end of 1826 4.37½
Fees to Land agent 2.00
Total — $26.37

July 22d 1825. Charles G.P. Hunt — Dr
To Cash $24.00
Augt 13th 1825. By 12 Chairs $32. Gold chain 31.50.
Ear rings $12. Coral Do $2.25. Slender $6.
Coral ear rings and armlets $9. Boyles acct 18.42
Total. $103.17
Balance due G.C.P. Hunt. — $79.17

Colonel Snelling copied the letter he wrote accepting
a duel with Lieutenant Baxley into his journal. *MNHS.*

Though some of his questions were disallowed, Hunter gradually shifted the focus of the hearing to Snelling's own behavior. Matters turned dramatically during the questioning of one particular witness. Captain Nathan Clark, one of the more respected officers in the 5th Regiment, was a settled, married man who avoided many of the petty quarrels among the officers. Lieutenant Hunter asked him, "Have you not heard Colonel Snelling say, a great number of times previous to the 31st July 1826, that if any of his officers were dissatisfied with the manner in which he conducted his command he was at all times ready to give them personal satisfaction?"

Snelling strongly objected to the question, rightly claiming that "it may implicate my conduct and discretion . . . but has no bearing on the case. . . . Justifying one's faults by those of another, it is bringing me before a tribunal instituted for the trial of Lieutenant Hunter when I, the prosecutor may be condemned on charges which ought to have been brought elsewhere." Hunter replied that he "had been led into an error by the advice and conduct of my prosecutor, and it is absolutely necessary for my defense that I should shew this to be the case." After a brief consideration, the court allowed the question. Captain Clark confirmed that he had "heard him (Colonel Snelling) make use of such or similar remarks." Though Colonel Snelling in his cross-examination tried to emphasize he meant his remarks to apply to "private injuries" not "official conduct," the damage was done.[43]

In his defense summation, given on October

21, Lieutenant Hunter restated in detail what had led him to make the challenge to Colonel Snelling, and much more besides, including the mention he had made in his note to the colonel of Snelling's invitations to challenges. He noted, quite tellingly, "I would ask any officer who had the least pretensions of honor or honesty, if he had received my note, and had not given me the invitations I there mention, if he would not have returned it and informed me I was mistaken." In effect, he was publicly accusing Snelling of setting a trap. After cataloging the insults and hardships of his long arrest, Hunter brought up others who had been insulted by the colonel's outbursts and the fact that the colonel had failed to return $2,000, "the long saving of a poor subaltern, which he has had in his possession for more than two years." Hunter concluded with a stinging attack on Snelling's character:

> I Saw with sorrow that those Gentlemen, who were the most ready to drink (with one whose duty it was to have set an example to his Regiment) to laugh at his obscene stories and stale jests, although repeated for the hundredth time, and to sacrifice the rights of their men, when his convenience or speculating disposition required it, were those who were patronised by our colonel.... And I soon found that in a Regt. supported by the people of the United States, the Laws by which we should have been governed were only enforced or rather used, by its commander on his enemies.[44]

Colonel Snelling very naturally asked the court to allow him the opportunity to reply to these serious assertions by Lieutenant Hunter, but Colonel Leavenworth's court in one terse sentence "decided against receiving replication" from Snelling. It was an insult to be denied a chance to refute Lieutenant Hunter's statements. But the trial was over.

The next day the court found Second Lieutenant David Hunter guilty of violating Article 25 and sentenced him to be cashiered from the army.

But the court, in its recommendation to President John Quincy Adams, who reviewed all such cases (and was himself a lawyer), strongly urged the president to set aside the sentence and restore the lieutenant to duty. Two months later, President Adams did just that. The president found Colonel Snelling's repeated offers to meet his subordinates in personal combat "subversive of all discipline ... and they disqualify to the common sense and feeling of mankind the officer thus self-degraded to the level of his inferiors from acting as a prosecutor against them for taking him at his word." But he also found Lieutenant Hunter's defense "full of irrelevance and abusive matter, much of which ought not to have been allowed by the court to appear upon the records: most especially as they denied to the prosecutor [Snelling] the liberty of replying to it. . . . The right of self-defense is sacred, but should not be suffered to be used as a cloak for slander." Still, President Adams remitted the sentence. In the end, David Hunter had fought his duel with Colonel Snelling, although with words, not weapons, and with the help of a sympathetic court, he had won.[45]

Soon after the trial, Colonel Snelling left St. Louis for Washington, DC. Though he was officially on furlough, much of his time at the capital was spent adjusting his commissary accounts as commandant of Fort Snelling. He seems to have kept poor records, insisting on controlling accounts but unable or unwilling to keep them satisfactorily, much to the irritation of the 5th Regiment's Assistant Commissary General Captain John Garland. The captain had been at loggerheads with Colonel Snelling from his arrival at Fort Snelling in 1826. Soon after the colonel departed the fort for St. Louis, Garland wrote Commissary General Jesup.

> Both from a sense of duty, and a desire to show the character of the officer under whom I have had the misfortune to serve at this Post, I submit the following statement for your private information. . . .

Quarter Master Sergeant Spaulding, a man over sixty years old, reported to me with tears in his eyes, that his colonel had deceived and ruined him, that he had borrowed of him about $740.00 and left the Post, without paying him, although solicited to do so in the most pressing terms. The sergeant has served all the active part of his life in the army and to use his own words, is left in his old age, a beggar.

It is notorious in the Regiment, that Col. Snelling, has thousands of dollars in his hands, the property of his junior officers, noncommissioned officers and privates, which he is unwilling or unable to pay. From this unprincipled course in his private affairs, what may we not expect of his public transactions?[46]

Not only does this letter support some of David Hunter's charges regarding Colonel Snelling, but it also indicates just how deeply disaffected many of the 5th Regiment's officers had become by the autumn of 1827. Colonel Snelling would certainly face a difficult task satisfying the commissary general's office.

Before the review of his books was well underway, however, Colonel Snelling's health took a decided turn for the worse. In November, his dysentery became very severe, and he was compelled to stay at the Washington home of his brother-in-law, Captain Thomas Hunt. The colonel extended his leave of absence because of his illness and did not begin the review of his accounts until the summer of 1828. Josiah Snelling died on August 20, 1828.[47]

⁓

So ended the career and life of Josiah Snelling. He was a complex figure, a man capable of rage or compassion toward his soldiers. But as an officer devoted to his duty, he demonstrated great energy and drive in building the fort that bore his name. A soldier of the old school, he was little concerned with "sashes and epaulettes" or the niceties of regulations. He committed violence against his own enlisted men behind closed doors, and he rented, bought, and sold other human beings. His tolerance, even encouragement, of "affairs of honor" among his officers undermined their discipline, while his lack of care or incompetence in financial matters, both the army's and his own, were disastrous to his career. Yet, when action was required or a decision needed, Colonel Snelling could act with alacrity—in fact, he seemed to relish the opportunity for any sort of military undertaking. In nineteenth-century military terms, he and the men of the 5th US Infantry succeeded in carrying out the difficult task of building and sustaining an impressive outpost far removed from any base of supply or support. In doing so, they were carrying out their government's ultimate aim of controlling and colonizing the ancestral lands of Dakota, Ojibwe, and other Native peoples in the Upper Mississippi region.

But if the Dakota had been unwilling to accept the post and the agency in their ancestral lands, building and maintaining the fort would have been much more difficult—if not impossible. From the time of Zebulon Pike's first visit, the Dakota had welcomed the newcomers, their goods, and the technology brought to them through the fur trade, and they viewed the fort as a conduit for that trade. When Fort Snelling was built, and for much of the decade following, the Dakota did not view the small garrison as a particular threat to their traditions and continued to consider the Americans' presence at Bdote as a loan of the use of the land. However, as much as officials like Lawrence Taliaferro wished to bring them peace and "civilization," the Dakota had little interest in abandoning their successful way of life. But the very trade they desired was beginning to erode that way of life, creating more dependence on the traders and ultimately the government for subsistence. Soon, government policies and changes in the economics of the fur trade would unravel the fabric of relationships and customs underpinning Dakota society.[48]

Edward Kirkbride Thomas, who served as a sergeant at Fort Snelling in the late 1840s and early 1850s, painted several views of the fort. *Oil, MNHS; gift of Abram Efelt.*

6. A Way of Life Unravels

FOLLOWING THE DEPARTURE OF COLONEL SNEL-
ling in October 1827, Major Josiah Vose held command of the fort. He took over a troubled garrison. From this time on through the following
year, as Taliaferro noted, increasing incidents of
bad behavior by the troops reflected their poor
morale and lack of discipline. The agent anticipated and perhaps knew in advance that the regiment's time at Fort Snelling was drawing to an
end. "The troops at this Post, Fort Snelling should
be removed without loss of time to the School of
Practice at Jefferson Barracks. 8 years is entirely
too long to keep the same troops at any one post.
The evil effects have been seriously felt, and are
daily and hourly experienced. Quarreling, tat-
tling, neglect of military duty, and a total absence
of all the finer feelings and of propriety and sense
of duty." The following May the 5th Infantry was
ordered to Jefferson Barracks in St. Louis.[1]

In May 1828 elements of the 1st Infantry took
over garrison duty at Fort Snelling and Fort Crawford. Between 1828 and 1848, the 5th Regiment
and the 1st Regiment rotated in and out of garrison duty in the "North West," with companies stationed at Fort Snelling, Fort Crawford, and Fort

Armstrong. This divided duty meant that only
a few companies occupied Fort Snelling at any
given time, depending on the local situation. For
about a year, the post was under the command of
Lieutenant Colonel Zachary Taylor. Colonel Taylor arrived at the post on May 24 and was present at Fort Snelling only a brief time, transferring
the 1st Regiment headquarters to Fort Crawford
in July 1829 to deal with the Black Hawk War.
He had yet to win his military reputation as "Old
Rough and Ready"—and later, the presidency. His
impact as a military leader was far greater in the
Black Hawk War than at Fort Snelling. Taylor, a
plantation owner who would own more than one
hundred enslaved people, brought two of them,
Jane and Glasgow, to serve him at the fort.[2]

Important changes were also underway in
Washington that would influence life at Bdote. In
1828, Andrew Jackson was elected president of the
United States. A former general and "westerner"—
he was resident of Tennessee, a relatively new state
west of the Appalachian Mountains—Jackson had
spent much of his earlier career fighting against
Indigenous nations in the South, and his policies
as president favored expansion at the expense

of Native peoples. These attitudes, which were shared by many whites, particularly those living west of the Appalachians, were put into action through the Indian Removal Act of 1830. This legislation authorized the government to exchange unspecified lands located west of the Mississippi River for the homelands held by the "Five Civilized Tribes"—members of the Cherokee, Muscogee (Creek), Seminole, Chickasaw, and Choctaw Nations—in what is now Georgia, Alabama, and Mississippi. A sum of $500,000 was appropriated for this purpose. The law resulted in the forced eviction of these nations from their ancestral lands and relocation onto unfamiliar territory in what is now Oklahoma. In addition to the land exchange, the dispossessed nations were guaranteed protection "against all interruption or disturbances from any other tribe or nation of Indians, or from any other person or persons whatever."[3]

The result of this law was the forced migration of tens of thousands of people from the southern United States onto the Great Plains. Jackson's secretary of state, Lewis Cass, the former governor of Michigan Territory, estimated that 93,500 "emigrants" would be added to the population of Native Americans already living west of the Mississippi, making a total of 244,870 in the region between the Mississippi River and the Rocky Mountains. The law also had the unintended effect of straining the already undermanned US Army, which could not transport or protect those being displaced. When the Seminole Nation of Florida resisted removal, the army was hard-pressed to force them out and suffered some bloody defeats during a decade of long, costly campaigns in the Seminole Wars. Given this strain on national resources, the 1st Infantry at Fort Snelling received few replacements or additional supplies to maintain the post.[4]

The removals also gave impetus to a further, final attempt to guarantee peace between the tribes of the Upper Mississippi region. Though they were not directly covered by the Removal Act, a number of tribes in the "Old Northwest" were affected by the policy behind it. The Potawatomi, Sauk and Meskwaki, Ottawa, and Shawnee would all be pressured into similar agreements in the years following the Removal Act. Further, it was now apparent to all concerned that the 1825 treaty had been a failure. As some of the Indigenous leaders had feared, the attempt to set fixed boundaries had not stopped conflict between the tribes and may have made matters even worse. In 1830, Secretary of Indian Affairs William Clark called for another meeting of the tribes of the Upper Mississippi region to attempt once more to make a final boundary settlement. The resulting fourth Treaty of Prairie du Chien, made among the Ojibwe, Dakota, and Sauk and Meskwaki, intended to set boundaries between these tribes and provided that each would give up some lands to create mutually recognized buffers or hunting zones between them. One of these areas specifically ran along the Mississippi from La Crosse to Prairie du Chien, then from the Mississippi to the Des Moines River in what today is southeastern Minnesota and northeastern Iowa; another was a large triangular tract of land in what would become southeastern Nebraska, northwestern Missouri, western Iowa, and southern Minnesota. Each tribe would cede lands along their mutual borders to the United States, which would guarantee that these buffer zones would remain neutral and unoccupied by either the tribes or white colonists. The 1830 treaty was the first since the 1805 agreement with Pike in which the Dakota ceded lands to the United States, though as in that treaty, the Dakota leaders and those of other tribes reserved the right to hunt on the ceded lands. This was also the first treaty in which the Dakota received an annual payment or annuity: $2,000 per year for ten years and access to a government-supplied blacksmith at the St. Peter's

Agency who would produce and repair guns, metal tools, and agricultural implements. Because the annuity was for tribal members only, the Dakota negotiated an additional consideration. The new treaty set aside lands on the western shore of Lake Pepin for "mixed-blood" or "half-breed" people—a largely European racial concept applied to people the Dakota referred to as "our relations." The idea was that individuals recognized as relatives by the Dakota, but not strictly members of the tribe by "blood," could be provided for. The reserve was never used, but the intended recipients of the lands were issued scrip, a document that entitled the holder to claim an equivalent amount of federal land anywhere in the country. This scrip was transferable, and much of it was sold to speculators, who soon created a very lucrative market for it.[5]

The treaty commissioners representing the United States, Clark and Colonel Willoughby Morgan, at that time in command of Fort Crawford, had a rather cynical view of the arrangement. In reporting to the secretary of war on the negotiations, they noted that the ceded tracts held excellent agricultural lands and showed good potential for mining. Their use for hunting by the resident tribes would last for only a few years.[6]

This new treaty would open the way for further and greater land cessions, a direct result of dramatic economic and social changes in the fur trade. In 1832, the business and political mastermind behind the American Fur Company, John Jacob Astor, decided to retire. He may have anticipated the coming decline of the fur trade, but there were others who were willing and even eager to carry it forward. Over the next two years, a reorganization of his business empire saw the company divided into two independent organizations. The former Western Department, which traded on the Missouri and its tributary rivers, became Pratte, Chouteau & Company, based in St. Louis, while the Northern Department, based

in New York, retained the American Fur Company name. Ramsay Crooks, a longtime power in the management of the AFC, emerged as leader of its namesake enterprise. A Scotsman by birth, Crooks could be personable and kindly in private life, but in the rough, ruthless world of the fur trade, he was renowned as a hard, shrewd, and canny business leader. In the words of historian Rhoda Gilman, Crooks was known for "his almost religious devotion to the advancement of the business under his direction." Soon after taking charge of the new AFC, Crooks began making changes. Among these was the decision to replace Alexis Bailly as the company's agent at Mendota.[7]

Bailly was an intelligent and very able businessman who had produced healthy profits for the AFC but had nonetheless become a growing liability. His contentious, sometimes vituperative clashes with officers at Fort Snelling, and especially with Indian Agent Lawrence Taliaferro over illegal importation of whiskey into the region, finally led the agent to cancel Bailly's license in 1834. There was no love lost between Taliaferro and the company, which had lobbied in Washington for years to have him replaced, but the Virginia aristocrat had friends in Congress as well, and he had managed to retain his post. Once Taliaferro canceled Bailly's license, however, consolidation became the AFC's best option to hold on to this still-lucrative concern. With this in mind, Crooks decided to reorganize the AFC's operations in the St. Peter's region, creating a new Western Outfit to be headquartered in Prairie du Chien under the management of three partners: the old company hands Joseph Rolette and Hercules Dousman, plus a rising and ambitious young clerk, Henry Hastings Sibley, who would be Bailly's replacement at Mendota.[8]

Sibley was only twenty-three at the time, but he had already acquired a solid reputation within the AFC, displaying a talent for business and personal dealings. His father, Solomon Sibley, was a

Agent Taliaferro formally canceled Alexis Bailly's license to trade on September 15, 1834: "I have declined to permit Alexis Bailly late an Indian trader within my agency to have further intercourse with the tribes of Sioux, not only in consequence of his *bad temper* but on account of his frequent violations of the intercourse laws, and particularly as developed on the 22nd of July last past." *Lawrence Taliaferro Papers, MNHS.*

recognition of Sibley's talent and a desire to keep him with the American Fur Company.[9]

Although furs remained plentiful, especially muskrat, the market for them steadily declined throughout the 1830s. In 1836, for example, Sibley shipped over 293,000 furs, hides, and buffalo robes from Mendota, but he still lost money. Smaller, independent traders like Philander Prescott, Joe Rolette, and Joseph R. Brown were already looking for other business possibilities. Prescott worked for Taliaferro, teaching European American farming techniques to the Dakota (discussed below); Rolette began illegally logging white pine timber on Ojibwe lands in the St. Croix River region; and Brown ran a dairy and livestock farm near Oliver's Grove (later Hastings), finding a ready market at Fort Snelling and among the civilians living on the military reserve.

Henry Hastings Sibley, about 1860. *Oil painting by John P. Bligh, MNHS.*

prominent attorney in Detroit who had served as a congressional delegate from Michigan territory. As a loyal Federalist, he was well connected with the political insiders of the Old Northwest. Henry, however, had no desire to follow his father into the legal profession. At the tender age of seventeen, with his father's reluctant blessing and assistance, he began working in various posts in the fur trade, first at Sault Ste. Marie and then at Mackinac. There he reported to Robert Stuart, another Scot, who like Crooks was an important figure in the AFC. From this position, Sibley worked for several years purchasing supplies for the various outfits, gaining an inside knowledge of the business and demonstrating his own abilities at negotiation and finding a bargain. Crooks offered him the position at Mendota in part as

These activities were a complete departure from the traditional attitudes of fur traders who, in the past, had always discouraged settlements and associated practices like lumbering and farming, both of which were viewed as detrimental to trapping and hunting. Henry Sibley was himself a departure from the old ways of the trade. As a well-connected white American Protestant, he was unlike the older traders. Though he would befriend Dakota people, join in on hunts, and even establish a relationship with Red Blanket Woman, the Dakota woman with whom he had a daughter, these kinship ties had less meaning or necessity for him than they had for his predecessors. In the end, his loyalty to the American Fur Company and his own career would come first.[10]

As the American Fur Company began to struggle with profits, it increased pressure on its tribal partners, charging higher prices for goods—more furs in exchange for a tin kettle, for example—and repossessing items given out on credit, like badly needed traps and guns. This practice was ultimately self-defeating, since it both inhibited the hunt and created a great deal of anger among the Dakota, who had seen such loans as gifts to kinsmen.[11]

In a meeting with Lawrence Taliaferro, the Dakota leader Black Dog complained of the changes:

> My Father, It is a matter of surprise to us that your Nation should come here among us poor Indians, to live hard & suffer upon this barren land. You quit a good country where you had plenty to eat and drink and to wear always good clothes. This seems strange to us. When the British used to see us we were better off, our game was plenty and our hunts good. They came but few among us & assisted us much. I do not say this that you are to understand me as disliking your nation—for we do not—but since your Nation has been here, they are catching at everything and times are altogether changed.[12]

Joseph R. Brown, about 1853. *MNHS.*

In the 1820s, Joseph R. Brown had been an enlisted man at Fort Snelling, but he was far more ambitious than most of his contemporaries. He worked to educate himself, curried favor with the Snelling family, and finally made an ideal marriage with Helen Dickson, the daughter of a Dakota woman and British trader Robert Dickson, who had dominated the fur trade among the Ojibwe and Dakota prior to the War of 1812. Three years after leaving the army in May 1828, Brown established a farm at Oliver's Grove, where he traded and also raised cattle, some of which he sold to the army. His operation grew to such an extent that it became a nuisance to the Mdewakanton. Wakiŋyaŋ Taŋka (Big Thunder, also known as Little Crow; he was the son of the Little Crow who signed the Pike treaty and the father of the war leader), complained that Brown

> has a farm & his cattle are roving over our lands and scaring off what deer [we] were in the habit of hunting & also fencing in our old encamping ground where always we place our Lodges to fish for our families. We wish him removed as his house is of no use to us in our trade. My Father—If he is not removed soon he will go on to make his house & fields larger & larger & cut our timber to sell to Steam Boats which he has done for some time & if it is of advantage to him it ought to be of some advantage to us also. But it seems he cares for nothing. We thought we would consult you as we cannot permit him there any longer.

Taliaferro duly wrote Brown, canceling his trading license for the coming year, but Brown would go on to make quite a career for himself. As early as the 1830s, he represented a new type of trader whose entrepreneurial drive and acquisitiveness would upset the centuries-old customs of the region and ultimately bring an end to the racially and culturally diverse community around Fort Snelling.[i] ꙅ

The combination of lower profits and a need for cash to pay its own creditors led the AFC to seek payment of Indian debts as part of treaty agreements. This solution gave the company an incentive to inflate the debts and encourage treaties. Where they had once opposed colonization of Indian lands as bad for the fur trade, they now saw the treaties as a source of hard cash. Fur traders were becoming Indian traders.[13]

———

For much of his time at Bdote, Lawrence Taliaferro had hoped to introduce European American agriculture to the Dakota. They had traditionally grown gardens of corn, beans, and squash in quantities sufficient to complement hunting and gathering. Women planted corn in hills in small plots. Though the corn crop was important, it was a supplement to other food sources, particularly game. Taliaferro hoped to establish what he saw as the more "civilized" practice of European and American agriculture, where men labored in the fields planting denser stands of corn in plowed furrows, producing a crop large enough to create a surplus that could be stored or sold for personal profit. Like virtually every white official, civilian or military, who had contact with the Dakota, the agent believed that Indigenous ways of life were doomed before the inexorable onslaught of Anglo-American "civilization" and its economic system. He saw European-style farming as the surest means of saving the Dakota and breaking their destructive dependence on the traders.

In 1829, Taliaferro's own relationships with the Mdewakanton improved. In May of that year, he recorded a peace dance held "by the relations of the young Sioux who were delivered up to the Cheppeways in May 1827, one hundred in number, at my house and at the Fort. This dance was to throw off their mourning." The following month, a council of Mdewakanton chiefs thanked him for supplying them with ammunition, since the traders had "taken away their guns." Taliaferro reassured them that the Ojibwe who continued to visit the agency received only tobacco, and not supplies intended for the Dakota. The traders' confiscation of guns and traps also affected the Wahpekute, whose chief complained to Taliaferro, "We are in a bad situation at this time. . . . We could not pay our traders and the consequence is that all our guns, traps, etc. have been taken from us. We are left to starve."[14]

This difficult situation led some Mdewakanton Dakota living near the agency to ask Taliaferro to help them to start farming as they had seen practiced at the fort. Though the gardens had not always enjoyed success, it is possible they influenced the Dakota. Cloud Man, the war chief of Black Dog Village, was willing to try it. He came to this decision after barely surviving a blizzard while hunting the previous winter on the Missouri River. Cloud Man and his followers, mostly from Black Dog Village, began farming in 1830 at a village known as Heyate Otuŋwe along the shore of Bde Maka Ska (named by the military for Secretary of War John C. Calhoun and known for many years afterward as Lake Calhoun). Taliaferro, who supplied seed and farm tools, also hired Philander Prescott to plow fields for the Dakota and teach farming methods. A few years later, in 1834, Samuel and Gideon Pond arrived, planning to bring Christianity to the Dakota; Taliaferro immediately hired them as agricultural instructors and helped them build a cabin near Heyate Otuŋwe.[15]

Referred to by Taliaferro as "Eatonville" (for Secretary of War John Eaton) or "my little Colony of Sioux agriculturalists," in keeping with his generally paternalistic view toward the Dakota, this Dakota farming community continued on part of the military reserve for nearly a decade. Like many aspects of society around Bdote in the 1830s, farming at Heyate Otuŋwe was something of a cultural compromise, rather than a wholesale adoption of European American practices.

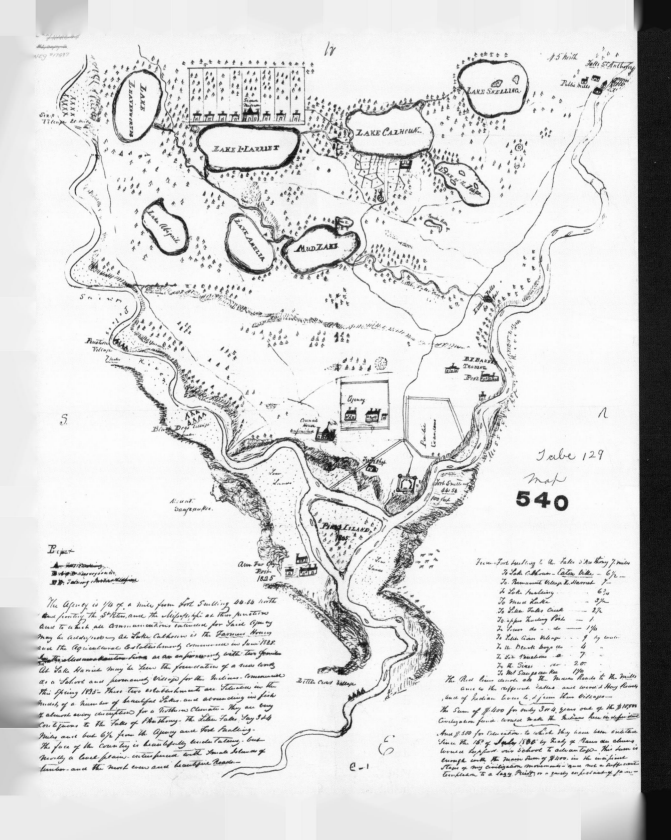

Table 129

Map

540

Women continued to do the bulk of the hoeing and planting on as many as fifty individual plots, and people of the village still made their customary winter hunting trips to the Rum River area. What seemed to have changed from the traditional Dakota practice was the scale of the gardens. According to Samuel Pond, they produced up to 2,300 bushels of corn. But unlike white farmers who might have stored the corn or sold it to the commissary officer at Fort Snelling, the villagers at Heyate Otuŋwe followed Dakota practice and shared their surplus with relatives in other villages.[16]

This proximity permitted another level of care from the European Americans, who also brought diseases. When an outbreak of smallpox occurred in the Upper Mississippi region in 1832, Taliaferro arranged with the post surgeon, Dr. Robert Wood, to have over five hundred Dakota at Bde Maka Ska and Kaposia vaccinated, a technique well known in the early nineteenth century and used among soldiers at the fort.[17]

While much of Taliaferro's attachment to Heyate Otuŋwe derived from his zeal to introduce agriculture to the Dakota, he had a strong personal attachment there as well. In 1827, he had married, after the custom of the country, one of Cloud Man's daughters, The Day Sets (Aŋpetu Inażiŋwiŋ), and the following year a daughter, Mary, was born from this union. So in part, the agent's support for the village was a way of providing for his daughter and his Dakota relatives.[18]

In fact, three other prominent men at the confluence married daughters of Cloud Man. Scottish American fur trader Daniel Lamont married Hushes the Night, who gave birth to their daughter Jane in about 1830. American Fur Company agent Henry Sibley married Red Blanket Woman, and their daughter Helen was born in 1841. Army Lieutenant Seth Eastman married Wakaninajin-win, or Stands Sacred, and their daughter, born in 1831, was named Wakantankawin, or Great Spirit

Woman; to the missionaries, she was known as Mary Nancy Eastman.

The European American women at the post apparently tolerated these liaisons, particularly when young, single officers were involved, but they were not willing to accept the children of these marriages into their society. All four men acknowledged their daughters and, to various degrees, tried to support them. Lamont died by 1837, but each of the others later married an Anglo-American wife who wanted nothing to do with her husband's Dakota offspring. Two of the daughters became part of the Anglo-American community. Jane Lamont was raised in the households of Gideon and Samuel Pond and married Samuel's nephew, Starr Titus, in 1850; she seems to have lived a culturally American life with him. In 1841, Sibley married Sarah Jane Steele, the sister of his business partner; his Dakota daughter Helen was raised in an Anglo-American household near St. Paul. Her background was well known, and Helen's prominent and popular father visited her and supported her. When she wed Dr. Sylvester Sawyer, in 1851, Henry Sibley attended the private ceremony to give his daughter away—and shared in his son-in-law's grief when Helen died of scarlet fever in 1852.[19]

Eastman and Taliaferro left the community to marry European American women, and their daughters retained their connections to the Dakota people. Taliaferro's daughter Mary spent her early years in Cloud Man's village with her mother, The Day Sets, and her grandmother, Red Cherry Woman. Her father had already gone back to Pennsylvania to marry Eliza Dillon, whom he brought to Fort Snelling for a time. Though Mary married a former soldier, Warren Woodbury, and lived in St. Paul with their children after his death, she eventually rejoined her Dakota family at the Santee Reservation in Nebraska in the 1880s. She remained there until her death in 1916. Nancy Eastman married Many Lightnings (Wakaŋhdi

Second Lieutenant Seth Eastman arrived at Fort Snelling in the spring of 1830, having joined the 1st Infantry at Fort Crawford the previous year as a recent graduate of the United States Military Academy. His first tour of duty at Bdote was short but eventful. He married Wakaninajinwin, or Stands Sacred, who gave birth to their daughter in 1831. That same year, the young lieutenant found himself in a serious, personal conflict with a number of Dakota. At the end of June, Ojibwe and Dakota leaders held councils at the fort relating to boundaries and peace agreements following the 1830 Treaty of Prairie du Chien. In the midst of one of these meetings, according to Lawrence Taliaferro, "Lt. Eastman without the least provocation" attacked one of the young Dakota men present, pulling him by the hair "in the presence of both nations, a great disgrace and affront." Relatives of the man came to his defense, and only the intervention of elders prevented them from killing Eastman. Though no official action was taken against the lieutenant, he did meet the man at Fort Snelling where, according to Taliaferro, "the matter was adjusted to the satisfaction of said Indian," probably through a gift or restitution of some kind.[ii]

Taliaferro remained annoyed over this disruption and distraction. Eastman's explanation for his behavior was that one of Penichon's sons had killed his hunting dog. On being questioned, this Dakota man said he had been drinking at the time and did not realize the dog was Eastman's. Taliaferro concluded, "If officers of the Army identify themselves with Indians and become as it were allied to their families—insults of the kind before alluded to may be expected sooner or later." The agent clearly believed that the killing of Eastman's dog and the lieutenant's reaction arose from a conflict between Eastman's Dakota in-laws and other nearby villages.[iii]

Lieutenant Colonel (later Brevet Brigadier General) Seth Eastman, about 1860. *MNHS.*

Ota) and lived with him near Redwood Falls; she died in 1858, giving birth to a son who would grow up to be known as Charles Eastman, a physician, writer, and activist working for Native American rights.[20]

Seth Eastman's skills as a draftsman were quickly recognized by his superiors, and he was assigned to a topographic project in Louisiana in 1832. According to family history, his parting from Stands Sacred and their little girl was heartfelt and tearful. But as far as is known, when he returned nearly ten years later with his second wife, Mary Henderson Eastman, the officer did not openly acknowledge or associate with his first family. In marrying his daughters to these Americans, Cloud Man was using Dakota custom to secure the well-being of his people through establishing kinship bonds. These men offered support for their Dakota families for a time, nodding to old customs. But unlike the French and English traders of previous decades, these American soldiers and businessmen had less interest in permanent, mutually beneficial bonds. They never actually lived with or "wintered over" among their Dakota relatives. As soon as their American wives were on the scene, social pressures from the white community required them to put their Dakota families in the background.[21]

In fact, the number of women and children at Heyate Otuŋwe suggests that there were more anonymous liaisons between soldiers at the fort and Dakota women at the village. What had been a mutually beneficial social arrangement providing kinship ties and community was becoming an abusive situation akin to prostitution. The Dakota custom of bride purchase, widely misunderstood among whites, devolved into what the soldiers saw as an exchange of goods for sexual favors from the young women at Heyate Otuŋwe. It is not clear how the women understood this situation.[22]

Heyate Otuŋwe was also connected to the early missionaries at Fort Snelling. Lawrence Taliaferro had sought to encourage the presence of missionaries "forming school of instruction for . . . this Nation." With the arrival of Gideon and Samuel Pond in 1834, his fondest wishes seemed to be met. The brothers were not ordained ministers, but Connecticut Congregationalists fired with the evangelical zeal of the Second Great Awakening. Though they were ardent Christians, the Ponds were amateur missionaries with a decidedly pragmatic outlook. They quickly realized that the Dakota were neither heathen savages nor noble primitives but people like themselves. If they were going to teach them agriculture and preach the gospel, the brothers would need to learn their language and understand their culture.[23]

Nonetheless, the Ponds shared the same view as Lawrence Taliaferro regarding the future of the Dakota: they must adopt Anglo-American civilization to survive the inevitable onslaught of "progress." In the long term, this would require European American–style education and conversion to Christianity, but in the near term they must learn to farm like white men. Taliaferro was impressed with the brothers and gave them his enthusiastic support:

> I am to furnish out of my private funds hay for the oxen belonging to the Indians and these young men are to have charge of them for the winter. They will plough some this fall and again in the spring for the Indians, and go thereafter to instruct them in the arts and habits of civilized life. . . . Some people cannot see how these two young men could leave their homes and friends to devote themselves to such a cause. . . . Laugh who will at these men, I have only to thank my God for permitting me to receive them.[24]

Though the Ponds had some success introducing agriculture at Heyate Otuŋwe, they made few religious converts. As other missionaries would also discover, Dakota people already had deeply held religious beliefs and a strong spiritual life. The missionaries failed to appreciate the reality that Dakota culture, shaped over centuries of successful adaptation, was complex and resilient.[25]

Within a few years of the Ponds' arrival, the American Board of Commissioners for Foreign Missions (or the ABCFM, as it was commonly known) decided to send out Jedediah Stevens, an ordained Congregational minister, to oversee the mission. As the ABCFM was the primary funder of Indian mission work among Protestant churches, the Pond brothers had little recourse but to accept this arrangement. Where the Ponds had respect for the Dakota as individuals, Stevens regarded them as "ragged, half-starved, indolent beings." He soon demoted the Pond brothers to worker status and rapidly undid whatever goodwill they had managed to create among Cloud Man's people. Stevens established his own school and mission at Lake Harriet that was attended by a number of children of mixed ancestry from Heyate Otuŋwe, but his mission lost most of its purpose when Cloud Man decided to move his village to the Minnesota River Valley in 1839. Increased tensions with the Ojibwe were supposedly the reason for the move, but Stevens's behavior and repeated problems with soldiers from the fort may have influenced the decision as well. The Pond brothers and their families followed the villagers to their new location, where they continued to practice larger-scale farming.

Though Cloud Man's efforts to emphasize agriculture at Heyate Otuŋwe might be viewed as a kind of assimilation, it was more likely a survival strategy, as hunting was becoming less reliable. Other groups of Dakota living near Bdote responded to the economic pressures from the traders in a more confrontational way. Though they could not afford to engage in open warfare with the traders, the Dakota helped to support themselves and send a message to their oppressors by killing and eating the traders' cattle and pigs. In July 1834, at the same time the Ponds were plowing land at Heyate Otuŋwe, men and boys from Little Crow's village, Kaposia, repeatedly killed cattle and hogs belonging to the Faribaults, openly consuming and drying the meat at their village. Though the traders complained to Agent Taliaferro, the latter had little sympathy for them, noting in his journal, "The Indians do not like these people, it is sure in consequence of their having had them as traders for so many years, and have been dealt hard with."[26]

Both the Heyate Otuŋwe and the Kaposia villagers were reacting to an increasingly difficult situation. Lawrence Taliaferro, though disgusted with the trader's tactics, saw no way to stop them, other than ending the dependence of the Dakota on their trade. He believed this could be accomplished through two new treaties, one with the Ojibwe regarding timber rights and the other with the Dakota providing for a cession of their lands east of the Mississippi.

Taliaferro hoped the Ojibwe treaty would

prevent serious conflicts between the tribe and whites who were engaged in illegal lumbering operations on Ojibwe lands. The rapid growth of American communities in the Mississippi River Valley during the 1830s resulted in high prices for timber. Some in the fur trade, who had access to markets, began cutting pine on Ojibwe lands in the St. Croix Valley. Joseph R. Brown began some timber operations as early as 1833, and in 1836, he had crews in the area of what is now Taylors Falls, cutting some 200,000 feet of pine logs. In March 1837, traders Henry Sibley, William Aitkin, and Lyman Warren tried to negotiate an agreement with the St. Croix and Snake River bands of Ojibwe to allow a ten-year timber lease on their lands, but opposition from Lawrence Taliaferro ended the attempt. At the same time, the Snake River band confronted another would-be lumberman, John Boyce, harassing his operations until he abandoned much of his equipment and retreated downriver. Other smaller operators quietly made their way into the St. Croix region, made gifts to the local bands of Ojibwe, and began cutting timber with little notice from the government.[27]

Taliaferro hoped that through a land cession, the Dakota would exchange part of their territory for a large, permanent annuity; the payment would end their dependence on the traders for subsistence and fund their transition to a farming economy. He was concerned, however, that the American Fur Company would interfere with the negotiation of these treaties. He believed the company's policies had aggravated the problems facing the tribes, and he feared the traders would try to claim exorbitant government payments for Indian debts, which the company would invariably exaggerate.

⁓

Negotiations for the Ojibwe treaty were held at Fort Snelling in the summer of 1837, and the results seemed to confirm the Indian agent's fears.

Land speculation in the United States, driven in part by the federal removal policy, had created an economic boom, driving the price of timber to new heights. Pressure was on to create a new treaty that would take pinelands from the Ojibwe. In charge of the negotiations was the governor of the newly created territory of Wisconsin, Henry Dodge. A ruthless speculator in lead mining and a shameless self-promoter, Dodge was an archetype of the frontier booster and exploiter. He was anxious not only to develop the white pine lumber in Ojibwe territory but also to see the Ho-Chunk people removed out of Wisconsin and beyond the Mississippi. In July 1837, the Ojibwe agreed to a treaty ceding their lands east of the Mississippi and south of Lake Superior to the United States. However, they were emphatic in reserving their right to continue using these lands for their traditional hunting and gathering. Their negotiators even specified which trees should not be cut down. The Ojibwe also received annuity payments, and $70,000 was set aside to pay off debts claimed by the traders.[28]

The treaty—and particularly the large payment to the traders for debts they claimed—was indicative of a substantial change underway in the region's economy. In the past, traders had resisted treaties that might encourage settlement at the expense of fur-producing habitat. But the decline in the fur trade, combined with an unsustainable credit system, made these debt payments a critical source of cash for the American Fur Company. As Thomas Jefferson had perceived, "Indian trade" would lead to debt, which would lead to acquisition of Indigenous lands, converting them to American property. In the words of historian Rhoda Gilman, "The elaborate charade of Indian treaties and land purchase was only the political and diplomatic window dressing necessary to disguise a subsidy to the fur companies for persuading or coercing their customer-employees to come docilely to the treaty table." Such payments became a standard feature in treaties with the tribes of the Upper Mississippi.[29]

In hopes of preventing the AFC from influencing the Dakota treaties, Taliaferro arranged for the negotiations to be held in Washington, DC, in August 1837. Twenty-six Mdewakanton, Wahpekute, Sisseton, and Wahpeton leaders agreed to make the trip, arriving in Washington in early September. Having witnessed the negotiation and signing of the Ojibwe treaty, they were probably aware of what would be proposed. It was a painful and difficult situation for the Dakota. Their dependence on the government and traders left them with little bargaining power, and the traders—who had followed in hot pursuit—were able to argue for substantial payments. The initial government offer was payment of $1 million for the Dakota lands east of the Mississippi River. The leaders initially hesitated to accept this offer, which they rightly pointed out broke down to "but little to each" for such valuable land. But the government commissioners refused to bargain, so reluctantly the Dakota leaders agreed to part with some of their ancestral lands in return for a permanent annuity.

All of Dakota territory east of the Mississippi, including the islands in the river, were given up for a trust fund that would pay them $15,000 in cash and $25,000 in food, goods, and farm implements each year for twenty years, as well as a payment of $110,000 for the "friends of the chiefs and braves"—their relatives of mixed ancestry. An additional $300,000 was to be invested by the US government in "safe and profitable State stocks," which would provide the Dakota approximately $10,000 per year "in perpetuity" to be used by tribal leaders and $5,000 to be spent at the direction of the president. There were real discrepancies in how the final version of the treaty was understood by the Dakota. The islands in the Mississippi included in the treaty lands were not mentioned by the Dakota chiefs at the negotiation,

MINNESOTA 1
SCALE.35 MILES TO 1 INCH

A. Hoen & Co. Lith. Baltimore.

A map created in the 1890s by Charles Royce shows Dakota and Ojibwe land cessions in what is now Minnesota. Lands ceded in 1825 by the Dakota are marked in purple; by the of 1851, with green marking the "half-breed tract" and pink, yellow, and blue showing the reservations along the Minnesota River that were taken after 1862. *Library of Congress.*

nor did they request any missionary schools, also funded in the treaty. These and other disputed aspects of the 1837 treaty eventually created an atmosphere of distrust and disappointment that would carry over for years.[30]

Initially, the Dakota treaty brought a feeling of optimism among the Mdewakanton that the promised payments and annuities would bring an end to their economic problems. Many even skipped their usual winter hunt, seeing no need to bring in furs. Although the treaty promised security and, by further separating the Dakota and Ojibwe, a measure of peace, the financial panic of 1837 made it difficult for the government to fulfill its promises, and Congress balked at approving the payments, leading to much unhappiness and suffering. As months dragged on, both Dakota and Ojibwe waited with growing impatience and desperation for the promised goods and payments.

Meanwhile, lumbermen and land speculators began to show up in the ceded lands, even before the treaties were ratified. By the following summer, whiskey sellers and even a few farmers were setting themselves up on the east bank of the Mississippi in the area of Bdote. One particular incident led Taliaferro to record a complaint from Chief Big Thunder: "I desire also to let you know that there are two men, Peter Perron [probably Pierre "Pig's Eye" Parrant] & old man [Abraham] Perry near us on the River. They have cattle & other property. Would it not be best to order them off, or until we hear from our treaty, tell them to be careful not to insult our young men. We look for these things, and our young men might do them some harm, or kill their cattle."[31]

⁓

For Fort Snelling itself, the cession of lands by the Dakota and the creation of Wisconsin Territory marked a significant change in status and mission. No longer an isolated garrison community in the midst of the Dakota homeland—"Indian territory"—the post was now on the edge of a region attracting more and more speculation and development. Like their Dakota neighbors, some of the civilian residents living on the military reserve around Fort Snelling experienced dislocation and economic hardship from the decline of the fur trade. Unlike the Dakota, however, some of these residents—along with some soldiers—would take part in that speculation. Even before the Washington treaty was ratified, both civilians and army personnel were pursuing property and speculative activities. Among the signers to the Ojibwe treaty were Samuel C. Stambaugh, sutler at Fort Snelling, and the post surgeon, Dr. John Emerson. Within weeks of signing the treaty, they were partners with another recent arrival, Franklin Steele, in the newly formed St. Croix Falls Lumber Company, and Stambaugh was petitioning the army to sell or lease to him the government sawmill at the Falls of St. Anthony.[32]

Stambaugh's activities are particularly informative. Like many other sutlers, he was a political appointee. He owned a Democratic newspaper in Pennsylvania and was an active voice for the party. His support for Andrew Jackson had been rewarded in 1830 with an appointment as Indian agent at Fort Howard (Green Bay, Wisconsin), but it had been an interim appointment that Congress did not confirm. He was compensated somewhat by being asked to serve as a secretary to a commission negotiating Indian removals to Arkansas in 1831. The Fort Snelling sutler post had come to him in 1835 as a further reward. Despite his title of "colonel," Stambaugh had no military or mercantile experience. Nonetheless, he clearly viewed the position as a potentially lucrative one. If not from selling goods and whiskey to soldiers, he might make his fortune by being in position to speculate on development opportunities ahead of others.[33]

From the time the sutler's post had come open, the American Fur Company had been trying to

The government flour mill and sawmill at the Falls of
St. Anthony, 1857. *MNHS.*

put one of their own in the job. The company
viewed all the army sutlers as a source of competition that they sought to eliminate or foreclose.
Stambaugh turned down several AFC attempts to
buy him out; only after he had visited Fort Snelling and experienced firsthand the challenges of
running a sutler's store did he agree to partner
with Henry Sibley.[34]

The partnership was part of Stambaugh's burgeoning relationship with the AFC. Early in 1836,
Stambaugh and fur trader Hercules Dousman met
with Taliaferro, "indicating as plainly as could be,"
noted the agent, "that I had better not do anything which would bring down the maledictions
of the Company Of Traders upon me."[35]

Long before the treaties were negotiated,
Stambaugh had expressed an interest in purchasing both the sawmill and the gristmill at the Falls
of St. Anthony. Writing to Quartermaster General
Jesup on February 17, 1836, the newly appointed

sutler stated, "It is generally conceded, I believe,
that mills are a great incumbrance to any frontier
military post," and offered to "improve" the mills
and supply lumber at a lower cost. When this
proposal went nowhere, he tried again, writing
Acting Quartermaster Colonel Trueman Cross in
September 1836 with an offer to supply the quartermaster at Fort Snelling "five thousand feet of
sawn plank" during each year of a twenty-year
lease. How and where he would acquire logs to
supply the mill he did not say, but as post sutler
he was doubtlessly aware of the demand for lumber and activities of traders like Joseph Rolette
and Joseph Brown, who were poaching timber on
Ojibwe lands.[36]

Though Stambaugh argued that the mill was of
little use to the government and in bad condition,
Acting Secretary of War C. A. Harris, when asked
to comment on the proposal, noted, "I do not
find, in the correspondence with this office, any

complaint of 'the expense and vexation of keeping up these mills,' as suggested by Mr. Stambaugh. The only proposition to dispose of them came from him." Indeed, the army still needed to produce lumber for repair of buildings and new construction at the fort. Though the War Department spurned Stambaugh's proposal, others would keep trying. In 1838, Henry Sibley, in partnership with two other fur traders, tried to obtain a lease, again to no avail. As more and more Americans moved into the ceded lands, there would be further proposals to acquire both the saw and grist mill. The Falls of St. Anthony presented an obvious site for waterpower development.[37]

Samuel Stambaugh was typical of the men seeking to take advantage of the "opening" of the lands east of the Mississippi. An easterner with no prior experience or association with the region, his connections were political rather than personal. His objective was to make money quickly by exploiting the region's resources. But with little practical understanding of the area and its challenges, outsiders like Stambaugh were at a disadvantage in competing with those already on the ground, who would eventually succeed where he failed.

⁓

While speculators in land and timber exasperated the Dakota, the opening of the eastern bank of the Mississippi also created serious headaches for the army. Excessive consumption of alcohol caused many problems in nineteenth-century American society; it reached its peak in 1830, when average consumption was 7.1 gallons of pure alcohol per year, though the efforts of temperance advocates saw this figure drop to two gallons by midcentury. The issue was particularly troublesome for the army. Soldiers received a daily ration of a gill (a half cup) of whiskey or rum, and post sutlers could sell each man—or woman, as civilians could purchase, too—up to two gills a day.[38]

Imbibers at Fort Snelling had an additional source in nearby fur trading posts. Traders could easily obtain bountiful supplies of spirits as rum and whiskey were distilled in various parts of the United States from New England to Kentucky. The traders were happy to supplement their income by selling to soldiers as well as Indigenous customers. Both the Indian agent and army officials had struggled since the opening of the post to control this source or cut it off altogether by threatening to withhold offending traders' licenses to operate in Indian territory. With the east bank and Mississippi islands no longer Indian territory after 1837, traders like Brown and Parrant quickly opened grog shops in the area. Brown's establishment, known as Rumtown, was opposite Camp Coldwater on the east bank of the Mississippi; Parrant set up shop at Fountain Cave. A veritable flood of whiskey was soon lapping around the edges of the military reserve, severely impacting discipline and morale at the fort.[39]

⁓

The fort's administrators continually faced other issues relating to civilians on the reservation. The gradually increasing population outside the fort and their growing herds of animals were creating problems and a source of conflict with the garrison for fuel and space. Major John Bliss, commander of the fort from 1833 to 1836, was a New Hampshire native and a veteran officer from the War of 1812 who had risen through the ranks in decades of frontier service. He seems to have been something of a disciplinarian, if not a martinet. Facing a number of discipline problems on his arrival, Bliss favored "the black hole"—a solitary confinement cell, without light—for particularly stubborn cases, according to his son's memoir. Offending soldiers were not the only inmates of the black hole. Major Bliss's enslaved servant Hannibal ran into trouble when he brewed "spruce beer" and sold it to soldiers. He was caught once

Through the 1830s and 1840s, many of the court-martial proceedings at Fort Snelling demonstrate the difficulties with liquor and desertion. For example, in February 1841, Private Edward Smith of I Company, 5th Infantry, was charged "In that he has been repeatedly guilty of drunkenness and riotous and disorderly conduct during the last six months as to render himself totally worthless to the United States." His punishment was "to serve one year at hard labor with a Ball and Chain on his leg, to have all pay stopped during that time . . . and to be allowed only such fatigue clothing as may be necessary for him."[iv]

Sometimes the lure of nearby liquor led to more serious charges of desertion. Desertion had become a significant difficulty for the army in general during the early nineteenth century. The incidents increased constantly until 1831, when some 1,450 men disappeared from the ranks, nearly one-quarter of the army's total strength. Such a serious situation prompted Congress to adopt new regulations. An 1833 act increased pay by one dollar per month, with some pay withheld until enlistees had served for a year. The term of enlistment was shortened from five years to three. These enticements were balanced by more severe punishments. Up to this time, most deserters were required to repay the government for the expense of their capture and then given a prison sentence, usually hard labor for a year. The new policy was decidedly draconian. Whipping was reintroduced, and some sentences called for as many as fifty lashes with a rawhide whip. This punishment was usually performed before the troops of the command, after which the prisoner's head was shaved, and he was dismissed from the service and marched off the post; Private Roster Nixon of Company F of the 1st Regiment received exactly this sentence when he was found guilty of desertion in August 1834. The decision of the court-martial was approved by General Gaines and likely carried out.[v]

There is no record of what happened to Roster Nixon thereafter, nor does his fate seem to have been a lasting deterrent. While desertions were not as common an offense as being drunk on duty, they continued to occur several times a year, regardless of the season. In December 1839 four men of the 5th Regiment were caught at a "Frenchman's house" or "Whiskey Shop" four miles downriver from the fort (probably the establishment of Pierre "Pig's Eye" Parrant at Fountain Cave, near where Randolph Avenue in St. Paul now intersects with the river). One of the men claimed "he was influenced by some of the party to go down to the whiskey shop after liquor, that he was aware he was doing wrong but that the idea of deserting never had entered his head." Though the court recommended leniency, the usual severe punishments were ordered, with the additional indignity of having the letter "D" tattooed on each man's hip. Interestingly, General Gaines seems to have had his fill of such extreme sentences. He reduced the punishment to the old penalty of forfeit of pay and time in the guardhouse, admonishing the officers, "regardless of the law in this description of punishments, to abstain in future from such decisions—decisions almost as discreditable to the Republic and to the age in which we live as if the eye, or the nose of the soldier were sentenced to be mutilated—decisions calculated to bring the rank and file of the Army into the deepest odium."[vi]

and punished, but the major's son recalled that a second offense "resulted in his catching a good licking and forty-eight hours confinement in the black hole, effecting a thorough reformation."[40]

Bliss had little sympathy for the civilian population around the post and a particular dislike for their cattle. He issued standing orders that any animals breaking into the post gardens be shot. This directive led to a particularly ugly incident in November 1834, as reported by the Indian agent: "Saturday 7th I find early this morning that the guard from the Fort have shot during last night several cattle belonging to persons at the Post (citizens). They had broken into the public field not until after the produce had been raised & put in store. There might be a few turnips & a few cabbage yet remaining on the ground but none of sufficient consequence to cause so many valuable work cattle to be shot by the soldiers." Major Bliss seems to have had an obsession about the gardens. He complained to Taliaferro about Dakota people fraternizing with the soldiers as they worked in the plots, and local "citizens" trespassing among them as well. The agent noted that the locals had been invited in to admire some "large cabbages, etc.," and agreed to tell the Dakota to "not even look" at the gardens or the soldiers. "Your soldiers alone are to blame," the agent replied. "If they will give the Indians will receive—but for the whites the Indians would do very well."[41]

The problem of scarce fuel involved the Dakota as well as the civilians. The army's woodcutting, added to the Dakota's use of wood and regular practice of using fire at the start of the fall hunt, had created an open landscape dominated by prairie around Fort Snelling and Bdote. By the 1830s, the army's consumption of wood began to interfere with the Dakota's need for this resource, adding to the general scarcity experienced from loss of game. In June 1836, for example, in council with "heads of villages," Lawrence Taliaferro recorded the Dakota's disappointment regarding lack of response from the army to their complaints that 3,600 cords of wood had been cut around their villages, destroying their "bark trees & sugar trees" (elms and maples) as well as the small timber needed for repairing or building summer homes.[42]

In general, interactions between soldiers of various ranks and the local Dakota seem to have increased during the 1830s, driven in part, at least, by the economic policies of the AFC. Indians who were unable to obtain credit from the traders in order to hunt turned to soldiers at the fort, and a surprising trade was developing. "Soldiers have been in the habit of loaning Indians their guns," Taliaferro noted in 1836. "Traps have been loaned to enable such to hunt as had not the means . . . the Traps which the soldiers in question procured to be made by a private Smith at this post, are loaned to certain Indians (unable to procure credits) on

shares. That is, the Indians give a portion of their hunt for the use of the traps." Though he saw nothing illegal about the practice, he feared soldiers might abuse their position, to the detriment of the Dakota and the licensed fur traders. He specifically noted an incident on April 17, 1836, where a soldier stole thirty muskrat pelts from an "Indian Woman," presumably Dakota. The agent was able to get compensation for her in the form of trade cloth, but this and other incidents made him uneasy. "The soldiers are too apt to impose on these people as these petty trespasses are not of infrequent occurrence," he wrote. "I shall apply to the commanding officer . . . [to] prevent the troops from trading in a small way with the Indians . . . as such proceedings may be carried too far and ultimately grow into an abuse & of a magnitude sufficient to interfere with the regular trade with the Indians of this country with their traders who give bonds and bear a license."[43]

There was a darker side to this interaction as well. Taliaferro made repeated remarks about soldiers harassing or assaulting Dakota women in nearby villages. For example, in July 1834: "Taylor of D. Co. (1st Inf.) and two other soldiers officially reported to Maj. Bliss for having beaten 3 Sioux women at their camp at Lands End [a trading post about three miles up the Minnesota River from Fort Snelling] on July 18, 1834. They were supposed to be drunk, as the men of the hay party are seen night and morning crossing the river at Black Dog's village and going down to Messrs. Faribaults and Bailly's trading house. . . . This in my mind is the way soldiers get and have gotten whiskey for some weeks past." The fort itself was also a source of whiskey for the local Dakota. Some commanding officers, like Major Bliss, denied Indigenous people any access to the fort, but others—particularly lower-ranking, temporary commanders—were more lenient, which created problems. Taliaferro complained in 1831, "Orders to pass Indians into the Fort, extended to officers generally. I have in vain applied to have all my Indians excluded as they get drunk in Fort Snelling and give me much unnecessary trouble."[44]

⸻

With the treaties of 1837, the army saw an opportunity and need to better define the boundaries of the military reserve and, at the same time, to address many of these problems. The limits of the reservation negotiated by Pike in 1805 had always been uncertain. According to Lawrence Taliaferro, the Dakota saw the reserve as the area one could view from the bluff where the fort was located, roughly two miles in all directions. The army had a more extensive approximation, based on the version of Pike's treaty as amended by Congress. This official version took in the Falls of St. Anthony and included land on both sides of the Mississippi, extending nine miles upstream from the fort. The Indian agent attributed the confusion to the treaty Colonel Leavenworth negotiated with the Dakota in 1820, which had never been ratified and "has since given us much trouble." The Dakota understanding of the Leavenworth treaty "included 4½ miles less than Pike—which oversight—has induced the Indians contiguous to this post to aver that they only gave Pike a mile around the present site of Fort Snelling."[45]

Around the time the 1837 treaties were being negotiated, the 5th Regiment returned to Fort Snelling, Fort Crawford, and Fort Armstrong, relieving the 1st Regiment. The 5th's commander was Major Joseph Plympton, who had been a senior officer at Fort Snelling under Colonel Snelling a decade earlier. Plympton had a far less tolerant attitude than his former commander toward the civilians who had been living on and near the reserve, holding them partially responsible for the excessive drinking by their sale of whiskey to the soldiers. A new survey, setting much larger boundaries for the reserve, would allow him to attack the problem by removing all civilians not connected to the army or the Indian agency as far as possible from the fort.[46]

The survey and mapping were undertaken by Captain E. Kirby Smith. He reported that there were 157 white residents on the reserve with no connection to the military, the agency, or the American Fur Company. (While people of mixed ancestry evidently did not merit counting, a missionary estimated a total population of four to five hundred in 1838.) Smith also estimated there were two hundred or more horses and cattle on the homesteads of these individuals.[47]

Following Major Plympton's instructions, the captain mapped out an impressive domain for the reserve. The border stretched north along the west bank of the Mississippi River to Nine Mile (now Bassett) Creek about a mile upstream from the government mills and Falls of St. Anthony, then angled back west through Lake of the Isles and Lake Calhoun, making a slight jog west of Lake Harriet to take in the mission there. Then the line ran a long way southeast to the St. Peter's River, upstream from Penichon Village, where it crossed the river. There it ran northeastward roughly parallel to the St. Peter's and the Mississippi to a point opposite Fountain Cave. Here the border turned northward, across the Mississippi and running beyond the bluffs to a point near the current line of St. Paul's Marshall Avenue, where it returned west to close the boundary on the Mississippi several miles downriver from St. Anthony Falls. The greatly enlarged reserve included not only Camp Coldwater and Cloud Man Village (both already regarded as part of the military reservation) but also Penichon's and Black Dog's villages. The American Fur Company post at Mendota, Joseph Brown's troublesome Rumtown, and Pierre "Pig's Eye" Parrant's establishment at Fountain Cave also fell within the new boundaries. With this claim, the army could rid itself of the nearby whiskey shops, control areas on the east side of the river with important wood resources,

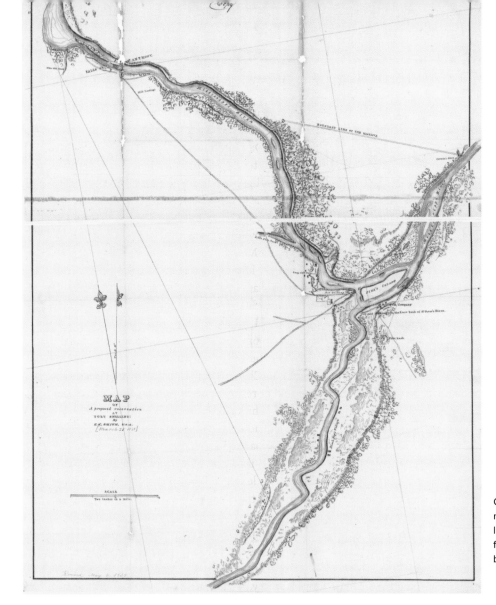

Captain E. Kirby Smith's map of the new Fort Snelling Reservation, 1838. A faint red line marks the boundaries. *MNHS*.

and keep out the anticipated flood of land-hungry colonists that would shortly arrive. Significantly, however, the new borders left a large area on the east bank of the river below the falls outside the military reserve, thus open for development.

Prior to the 1837 treaties, some of the settlers who had lived and farmed in the reserve for a decade and more became concerned for their future in the area. Their clashes with Major Bliss in 1835 had left them feeling insecure about their holdings. The army's attitude toward their presence had clearly changed since Colonel Snelling's time. In an 1836 letter to the Department

of Indian Affairs, Lawrence Taliaferro reported his views on the settlers in response to a petition to President Jackson from some of them seeking official permission to remain on their farms. At that time, Taliaferro replied that his agency "had not been seriously incommoded" by the presence of the civilians and noted that they had been, to date, less troublesome than some of the licensed fur traders. He also pointed out that many of those living within the reserve had personal relationships with the Dakota and Ojibwe, "as nearly the whole of said petitioners are half bloods of the Sioux and Chippewa tribes, and remain . . . allied to those tribes by marriage ties." By residing near the fort, they were seeking "more security from occasional . . . war parties." Still, he recognized the army's concern, "that the military post here will suffer a loss in fuel, and otherwise be interfered with by the ranges of private cattle upon the lands continuous to it and by the formation of enclosures." Taliaferro concluded by suggesting that all the residents "not connected officially with the government" be moved "to the East of the Mississippi . . . where I am disposed to believe that they will be in the way of no one."[48]

No action was taken on the petition to Jackson. Still feeling they had some claim to their farms, and with the encouragement of Sibley and Stambaugh, for whom they were valued customers, a number of residents petitioned newly elected President Martin Van Buren in August 1837 for relief: "The undersigned will further state that they have erected houses and cultivated fields . . . and several of them have large families of children who have no other homes." Among the signers were Louis Massie, Abraham Perry, Peter Quinn, Antoine Pepin, Duncan Graham, Jacob Fahlstrom, Oliver Cratte, and Joseph Bisson. Samuel Stambaugh was asked to support the petition and later wrote both President Van Buren and Secretary of War Joel Poinsett, defending the settlers' position. As sutler, he was also defending his customers, more valuable than ever because the Seminole War had seen the Fort Snelling garrison diminished.[49]

The petition was once again ignored and made somewhat irrelevant by ratification of the 1837 treaty, after which a number of the residents relocated as Taliaferro had suggested to the east side of the Mississippi. Within a few years, Taliaferro's views regarding civilians in the reserve would change, as some of them became more involved in whiskey selling and illegal trading. By the time of Major Plympton's redrawing of the boundaries, some had already moved.[50]

And what did the Mdewakanton make of the new boundaries? All through the 1820s and 1830s, the Dakota continued their lives in the area around Bdote. Taliaferro's diaries note gatherings, dances, ball games, and initiations in the area of the reserve. They had signed no new agreements; the military was permitted to live in the area, but it did not "own" the land. There was no reason to believe the Mdewakanton had changed their perspective. At least one commander of Fort Snelling agreed. In an interesting letter written in May 1837, near the time the Ojibwe treaty was being discussed, Lieutenant Colonel William Davenport, then commander of the fort, stated his belief, based on the Pike treaty and the 1830 Treaty of Prairie du Chien, that "we have by both of them, nothing more than the consent of the Indians to the establishment of a military post here. . . . We are upon Indian land, and cannot consequently, grant any part of it to any person whatever for his private use." Though he was discussing the proposed sale of the government mills, he clearly understood the reserve west of the Mississippi to be Dakota land.[51]

As 1837 and 1838 wore on with no word that the treaty had been ratified and no sign of promised payments or supplies, the Dakota became increasingly uneasy, believing they had been misrepresented and cheated. Economic conditions had not improved, and scarcity of game made their situation ever more difficult. Traders, unable to sell muskrat furs, refused to extend more credit. Consequently, some Dakota killed livestock belonging to the agency, traders, and civilians, sometimes out of anger, often for food. Taliaferro recorded Big Thunder's frank explanation: "If Indians were killing horses and cattle it was not surprising." He felt sorry for it, but it was all brought by actual starvation: "My father, the severe disappointment in not getting our goods & money early this spring under our treaty of last year has rendered our people miserable. Our Traders have stopped credits, our game is gone and a man may starve one and two days and even three but on the fourth he becomes desperate and kills the first thing that crosses his path. Hence is it surprising our people occasionally kill an ox or a horse?"[52]

"Colonel" Stambaugh was replaced as Fort Snelling's sutler, first by Benjamin Baker, the long-time trader and resident at Camp Coldwater, and following his death in the spring of 1839 by Franklin Steele, one of the more dynamic and controversial figures in the region's history. Born in 1813 to the family of General James Steele, a prominent figure in Pennsylvania military and political affairs, he had through his career benefited from strong Democratic party connections. Family tradition relates that young Franklin was encouraged to seek his future in the Fort Snelling area by no less a person than President Andrew Jackson. He seems to have arrived at the fort in the summer of 1837. Following a largely unsuccessful attempt to found a lumber milling company on the St. Croix, he sold out his shares and—probably through political connections—took over the sutler post at Fort Snelling.[53]

Where Stambaugh had failed in dealing with the army and American Fur Company or making much headway as sutler, Franklin Steele was a dynamo. Outgoing and persuasive, the young Pennsylvanian was soon making deals and connections within the army community. Steele had no qualms about working with the American Fur Company and quickly formed a partnership and friendship with Henry Hasting Sibley. He was well suited to take advantage of his position as

Franklin Steele, about 1856. *MNHS.*

sutler. Though most of his army customers purchased on credit, the sutler was entitled to a place at the army paymaster's table when the troops received their pay. The government paid its soldiers in cash, so Steele was able to accrue hard currency, an advantage to anyone doing business on the frontier, where bank notes and drafts were often suspect. A cash balance also allowed him to make quick, advantageous transactions, particularly when purchasing land. Steele's position at the fort also gave him access to information. The mail was delivered there first, so people in the garrison were often the first to know about outside news.[54]

Steele was certainly aware of the new boundaries of the military reserve and the notable fact that the east bank of the Mississippi below the Falls of St. Anthony was not included within its bounds. As soon as the 1837 treaty with the Dakota was approved, this stretch of land—the best available for potential waterpower development—could be obtained by whoever first staked a claim.

In December 1838, Steele's political contacts informed him that the treaty had been ratified. He had this valuable information in the same mail that his potential rivals for the land had received it. This was a group of officers, including Major Joseph Plympton and Captain Martin Scott, who were actively seeking to acquire land in the newly ceded territory. Steele learned that the officers planned to go out early the following day to lay out a claim. The sutler, along with his sometimes business partner Norman Kittson, decided to go that very night. They loaded a wagon with supplies and crossed the frozen river in the dark. Overnight, they staked out their property, set up a rude shack, and "planted" potatoes in the snow, thus making the necessary "improvements" to secure their claim. The story goes that they greeted the disappointed officers the next morning by offering them breakfast at the new "farm."[55]

In the next few years, Steele claimed or bought nearly a mile of riverfront property below the falls, but it would be more than a decade before any of the area would be developed for milling. The financial depression of 1837 slowed development, but the size and location of the military reserve also created problems. It was as if the reserve's boundaries were set in order to discourage development. All of the good landing places for steamboats in the area around Bdote now fell within its borders. Both Steele and Stambaugh, who continued his quest to purchase the government mills, objected to the extent of the new boundaries. They were joined by others, notably Joseph Brown, who had expected a new city to grow across the Mississippi from the fort. He lost his establishment at Rumtown and a number of land claims around it. Brown's lobbying persuaded the legislature of Wisconsin Territory, whose boundary now extended to the east bank of the Mississippi, to protest the expansion of the reserve.[56]

All of these parties were convinced that the extension of the military reserve east of the river had more to do with the speculative plans of army officers at Fort Snelling than the well-being of the fort's garrison. Major Plympton's attempt to claim lands at the Falls of St. Anthony seemed to confirm this notion. Meanwhile, another group of officers hired Philander Prescott to take charge of a claim they planned for a townsite near the mouth of the St. Croix River, offering him $1,000 and a one-eighth share; the town also took his name. Agent

Taliaferro described the situation as a "rage for speculation," presumably among both officers and traders, though he appears not to have been immune from it himself. He complained, "It was my intention to take up a small tract at the mouth of the St. Croix, and went up it for this purpose, but while absent on duty with my Delegation [negotiating the treaty] at Washington, The officers of the 5th Regt. Stationed at Fort Snelling laid their hands upon it. . . . So old Residents have been thus defeated in a place, after passing the prime of their lives in the Indian country."[57]

Despite the protests, the secretary of war accepted Major Plympton's expansion of the reserve's boundaries. In October 1839, he ordered all of the civilians off the reserve, including some who had moved to the east bank of the river. They were given until May 6, 1840, to move. According to the US marshal for Wisconsin, the settlers had initially complied with the order, but, as he testified later to a congressional committee, "within a day or two of the period assigned for their removal (6th of May,) they evinced a contrary disposition; and finally, their determination appearing decided, resort was had to military force. . . . The next day (May 7) the reservation east of the river was entirely clear of intruders. The greater part of them being Swiss, the services of an interpreter were required. To these circumstances may be partially attributed the necessity of using force." The force involved was a detachment of troops sent from Fort Snelling by Major Plympton at the marshal's request; they physically removed the last residents and burned their buildings.[58]

Most of those ordered off the reserve had little in the way of property. They quickly found a convenient location just outside the military reservation, ten miles downstream on the "east" side of the Mississippi where the river makes a wide turn to the north. Here, at the appealing place just vacated by Little Crow's community, was a firm, sandy riverbank good for landing canoes and, in

time, steamboats. Several small brooks ran down from the high ground to the north, offering good water and open areas for gardens and grazing.[59]

Among the first squatters to relocate there was the voyageur-turned-whiskey-trader Pierre Parrant, known as Pig's Eye because he had a malformed eye. He moved from Fountain Cave and sold "ardent" spirits to soldiers and Indigenous customers alike from his shack. The little cluster of cabins and holdings that grew up around "Pig's Eye's landing" between 1838 and 1840 included some of the same multiracial trading and farming families who had originally inhabited the area around Coldwater Spring. They had no intention of founding a village, much less a town or commercial center. It was a singular instance of unintended consequences, however, that Major Plympton's expansion of the Fort Snelling reservation meant these unassuming folk were staking out claims at the head of navigation on the Mississippi River, a strategic spot for any future development in the entire region. The inhabitants of Pig's Eye carried on much as they always had. The community traded property among themselves almost like a form of currency: so many acres for a horse, a half claim for a wagon or just as a gift to attract a new neighbor. But others on the scene cared much more about the economic potential of the site. Soon Henry Sibley, Norman Kittson, and Franklin Steele were paying cash for claims around Pig's Eye. It would take some time yet, but the beginnings of St. Paul were well underway.[60]

If the Dakota were unhappy, so was their agent. Lawrence Taliaferro had gambled that the resources promised to the Dakota in exchange for their lands could be directed toward the conversion of their economy to agriculture and "civilization," breaking their dependence on the traders. In this, he simultaneously overestimated the willingness of the Dakota to change their entire way of life and underestimated the ability of the AFC agents to hang on to their Indian "assets" and the government cash they generated. The long delay of goods and annuity payments under the 1837 treaty had diminished his standing with the Dakota. The benefits he had hoped would come from the agreement failed to materialize, while the influence of the AFC and its representatives among the Indians seemed as strong as ever.

In his nineteen years of service, Taliaferro had shown himself to be a man of contradictions. A slave owner with a southern patrician's sense of superiority to all beneath his race and his station, he acted with what he believed was a detached policy in his dealings with traders and his Indian "children." Forthrightly paternalistic, Taliaferro

The confluence of the Minnesota and Mississippi Rivers, about 1851; see page 118. *Detail of oil by anonymous artist, MNHS; gift of Russell Blakely.*

firmly believed that the US government's Indian policy, if followed honestly and honorably, was benign and represented the only secure future for Native people. Over time, he became convinced that the main impediments to realizing this policy were the traders. In a report to his superiors written on July 4, 1838, he noted, "The cupidity of the trade created a fixed determination. . . . [A] monopoly by one company [the AFC] necessarily produced the desire to frame and use any species of council, or artifice and cunning—no matter who suffers—whether individual or government—to the end that no one not even the official agents of the United States should compete with them in their fixed determination of securing and holding an influence over the Red Man, which all the forces [?] of the laws of heaven . . . or justice should not shake."[61]

Through years of contact through councils, gift giving, and his own marriage, and aided by the remarkable diplomacy of his interpreter Scott Campbell, Taliaferro had gained a degree of trust from the leaders of local bands. In turn, he had become increasingly sympathetic to their needs. Beginning about 1830, his journals show that his speeches in council with both Dakota and Ojibwe leaders often begin with the words "My friends," rather than the previous "My Children." Seventeen years after taking his post, the agent would reflect in his journal, "I have succeeded in all

I desired with these people [the Dakota] without resorting to force or calling on the military for aid."[62]

In the summer of 1839—disappointed, embittered, and wearied of the continuing conflicts with traders—Taliaferro resigned. Reflecting on the end of his career some twenty-five years later, he wrote (referring to himself in the third person), "This was well, for he well knew that the time would come when all his efforts to do good would pass into oblivion."[63]

Both the army and the Dakota would soon miss the agent's presence. Taliaferro's successors were also political appointees, but they had strong ties to speculators and little interest in the Dakota or their future. They would only accelerate the deterioration of the Dakota's situation. According to the agent's own recollections, Little Crow—the grandson of the chief who signed the 1805 treaty—visited Taliaferro's Washington office when he and other Dakota leaders were in the city for the negotiation of the 1858 treaty. After expressing his own affection for the former agent, Little Crow lamented, "How is it. You counciled our nation for more than twenty-one years and since you left we have had five agents. . . . We failed to get a friend in anyone like you: they all joined the traders."[64]

Fort Snelling from Mendota, about 1850. *Oil by Henry Lewis. MNHS; gift of William Cutler.*

7. An Outpost and Source of Change

WHILE TREATIES WROUGHT CHANGES IN THE lands bordering Fort Snelling, a change in transportation aided the colonization of the region by lessening its isolation from eastern states and making it a destination for a wider range of American citizens beyond fur traders and explorers. With improvements in design, engine power, and river navigation, steamboats became the preferred method of transporting supplies and people to Fort Snelling and the nearby AFC outposts, and soon to the new towns of the region.

By 1833, around twenty-five boats a year were calling at the fort, carrying with them more and more visitors arriving as tourists. These included artists like George Catlin, who returned East to exhibit his romantic landscapes and portraits of Native people, promoting the area as a destination for "the fashionable tour." Travelers who acted quickly might see the beauties of the region and its Native population before what Catlin called "the grand and irresistible march of civilization" transformed them and swept them away.[1]

Well-appointed steamboats called at the fort, where a regimental band might entertain the visitors while officers and ladies from the garrison led them on a picnic to the "Little Falls" (Minnehaha Falls) or the Falls of St. Anthony. Some visitors might be entertained by a buffalo dance by the "joint bands of Sioux" in exchange for a few barrels of flour and crackers from the boat's supplies. This exploitive sort of tourism was welcomed by the Anglo community at Fort Snelling. It connected them to their own world in the East and countered somewhat the reputation of the region as a frozen "Siberia" unfit for agriculture or commerce. Indeed, Taliaferro observed in 1835 that the climate in the area was "salubrious . . . for six or seven months of the year," had "fine scenery, beautiful country, excellent water, incomparable climate for general health and an open field for sports men." In fact, all manner of visitors would be welcome except "black legs [cheating gamblers], steam doctors [quacks], pick pockets, abolitionists, kid nappers and whiskey introducers." Wealthy travelers from much of the United States made the journey. Easterners came to see the exotic West, while southern plantation owners (with enslaved servants in tow) found the region a welcome respite from the heat of a southern summer; both groups were ready to check out its business prospects.[2]

Fort Snelling was the head of navigation on the Mississippi. A steamboat is docked at the landing. *Watercolor by Seth Eastman, 1848, MNHS.*

In the summer of 1823, the steamboat *Virginia* was the first to puff its way upstream to the landing below Fort Snelling. Its arrival "became an era from which we reckoned," recalled Charlotte Van Cleve. Though other boats would follow, they were initially few in number. Their main business so far north consisted of hauling supplies and occasional passengers for the government; most of the commerce lay farther downstream, near the lead mines of Galena.[i]

The design of the boats themselves, and the nature of the river, created other problems in reaching the Upper Mississippi. In the 1820s, most steamboats were essentially keelboats with a steam engine running a stern-mounted paddle wheel. They did not have enough power to run upstream during periods of high water, and their draft was too deep for effective navigation in times of low water. These were particular problems when dealing with the river rapids at Rock Island and the mouth of the Des Moines River. By the end of the decade, newer designs based on the wider, shallower-draft flatboat hull came into service. These improvements, along with more powerful engines and boilers, gave the river steamboats more cargo capacity and ability to deal with a variety of river conditions. In the late 1830s, Congress authorized US Army engineers to make the first navigational improvements on the Upper Mississippi at the Des Moines and Rock Island rapids. Led by a promising young captain named Robert E. Lee, a crew of soldiers and civilians blasted and dredged a wider, straighter channel through the bedrock, greatly easing passage through the area. But no advancement in boat design or enhancement to navigation could overcome winter. Whenever ice covered the Mississippi, particularly at Lake Pepin, all river travel came to a halt, and the accustomed isolation returned. Only the building of railroad connections in the years following the Civil War would change this reality.[ii]

Taliaferro's list of unwelcome guests suggests that the Indian agent might not have appreciated one of the newer arrivals at Fort Snelling. As part of the Army Bill of 1838 that attempted to improve life for enlisted men, each post was allowed to have a chaplain. Accordingly, the Council of Administration at the fort elected to call the Reverend Ezekiel C. Gear for this duty. Reverend Gear was an Episcopal clergyman, living at that time in Galena, Illinois; his son-in-law, Lieutenant Samuel Whitehorne, was then serving with the 5th Regiment at Fort Snelling. Though these connections recommended him to the garrison, the fact that he was an abolitionist may have given some officers, and Major Taliaferro, pause. But his beliefs regarding slavery seemed to have mattered little to the garrison at Fort Snelling, where Reverend Gear would serve for twenty years as chaplain and sometimes schoolmaster—and the practice of slaveholding continued.[3]

Not long after Lawrence Taliaferro left his post, the small polyglot settlement at Pig's Eye gradually began attracting residents. Soldiers recently discharged from the army, particularly Irish immigrants, found it an appealing place, relatively free of the social prejudice of eastern cities. The presence of French/Ojibwe/Dakota families, Swiss refugees from the Red River Colony, and even a small number of freed Blacks, like James Thompson, preserved the openness and ethnic fluidity that had existed in the area around Fort Snelling. In 1836, the region east of the Mississippi became Crawford County, a part of Wisconsin Territory, with the ubiquitous Joseph R. Brown serving as justice of the peace. He was also the representative at the territorial legislature in far-off Madison. Two years later, when Iowa became a territory, Fort Snelling and the military reserve west of the river, still Dakota land, fell under its jurisdiction as Clayton County; Henry Sibley served as justice of the peace. So by 1840, the region surrounding Fort Snelling and Bdote came under US civil authority. Though most of the area that was nominally Iowa Territory is part of the Dakota homeland, the political pressure to force the Dakota to sign another treaty ceding the land was continuous and growing.[4]

As a representative for the American Fur Company, Henry Sibley had spent much of 1842 in Washington, DC, lobbying Congress to pass the Doty Treaty (see sidebar page 112). The AFC's financial situation had deteriorated steadily after 1837, and the company looked to the treaty as a possible cash infusion. The treaty died in a final Senate vote in August 1842, and the AFC soon closed its doors for good. For Sibley, the journey east was a business failure but a personal boon. In April 1842, he attended the Baltimore wedding of his business partner and friend Franklin Steele. There he met Steele's younger sister, Sarah Jane, who was to travel west with her brother and his new bride, Anna Barney, and live with them. Over the following winter, Sibley courted Sarah Jane Steele at Fort Snelling. On May 2, 1843, Sarah Jane Steele and Henry Hastings Sibley were married at Fort Snelling by Reverend Gear. A business alliance thus became a matter of family as well—again.[5]

The early speculation at Pig's Eye blossomed with the beginnings of commerce as steamboats began calling at its convenient landing spot on their way upriver to the fort and the fur company

A fife owned by Joseph R. Brown in the 1850s. *MNHS.*

In the first part of the nineteenth century, bands were not an official part of regimental organizations, but in many US Army regiments, the traditional instruments of bugle, fife, and drum were augmented with other brass and woodwinds to create a band. Though there was no official provision for the extra instruments, they could be purchased by the post's Council of Administration. This was the same group of officers that regulated the sutler's fees and purchased books and periodicals for the post library. In fact, the regulations governing the councils specifically allowed that funds could be used for the "maintenance of a post band."[iii]

Colonel Snelling gave instructions for parades and reviews that included martial music. He requisitioned musicians' coats from the quartermaster stores in St. Louis, a request that was duly filled.[iv]

post at Mendota. The town also became the southern terminus of a nexus of trails between Fort Gary in the former Selkirk Colony and head of navigation on the Mississippi River. Since the 1820s, there had been trade and travel between the colony and Fort Snelling. By the mid-1840s, caravans of oxcarts driven by Métis traders bearing loads of furs, mostly buffalo robes, made their way from the Red River Valley to the new settlement. Outsiders from other parts of the United States came as well, not only as sightseers but as residents. Through the early 1840s, the village became a small town, and Pig's Eye became known as St. Paul. This renaming was the suggestion of Catholic missionary Father Lucien Galtier, who founded a chapel by the same name on the bluff above the landing. Along the St. Croix

In 1841, the governor of Wisconsin Territory, James Duane Doty, negotiated a treaty with a number of Dakota chiefs that would have created an Indian territory in what is now southern Minnesota. This region would be under government control and supervision, with the stated purpose of establishing agricultural communities among several tribes, including the Ho-Chunk and Sauk and Meskwaki. One of the more remarkable provisions of the Doty Treaty, as it was known, would have allowed qualified Native Americans to become full US citizens.

The treaty won support from Iowa and Wisconsin politicians, anxious to see their Native populations removed to this new Indian territory, as well as the various traders who hoped to profit from the ensuing contracts and annuities. The newly elected president, William Henry Harrison, also favored it. But Harrison died soon after taking office, and senators from slave states, notably Thomas Hart Benton of Missouri, vehemently opposed the idea of granting citizenship to any people of color, so Doty's treaty failed to gain enough support for ratification.[v]

River, the logging community of Stillwater was also attracting settlers. Although by no means a massive immigration, these new colonies were becoming large enough that the idea of creating a Minnesota Territory became increasingly appealing. This had, in fact, been the intent of Illinois Senator Stephen Douglas in setting the boundaries of Wisconsin and Iowa.[6]

Locally, the effort was led by former traders and speculators like Henry Sibley, Joseph Brown, and Franklin Steele, who stood to gain from the increased value of their holdings and the potential political rewards of new federal and territorial offices. But it was Sibley who came to the fore. Politically experienced and widely trusted by citizens in the nascent towns of Stillwater and St. Paul, Sibley became the chief lobbyist for territorial status following the Stillwater Convention of August 26, 1848. Arriving in Washington, DC, that autumn, he found a strong ally and supporter for the cause in fellow Democratic senator Stephen Douglas of Illinois. Though not a voting member, the eloquent and knowledgeable Sibley impressed the congressmen, who had expected a backwoodsman in buckskins. With much help from Douglas, a bill creating Minnesota Territory wound its way through partisan and sectional complications and was signed into law on March 3, 1849.[7]

The creation of Minnesota Territory would have significant consequences for the Dakota and Ojibwe people of the Upper Mississippi region and for the Fort Snelling garrison. Those who lobbied and campaigned for its establishment knew well that the original boundaries between the Mississippi and St. Croix Rivers were insufficient for the creation of a new state. Only the acquisition of the remaining Dakota and Ojibwe lands by the United States would make statehood possible, and this became the primary objective of nearly all politicians and boosters in the new territory. For the garrison at the fort, the growing population of Minnesota Territory would renew

Red River oxcarts on Cedar and Third Streets in St. Paul, about 1860. *Photo by Charles A. Zimmerman, MNHS.*

pressure to reduce the size of the military reservation and eventually bring into question the very need for the post's existence.

Starting in the mid-1830s, the need for troops for the Seminole War in far-off Florida meant companies were transferred out of Fort Snelling and other western garrisons, and new recruits were sent elsewhere. By September 1841, only two companies of the 5th Infantry remained to be relieved by Companies D, G, and H of the 1st Infantry. With them, in command of Company D and acting command at Fort Snelling, was Captain Seth Eastman, now returning to the confluence. He had enjoyed a long assignment at the United States Military Academy, where he had instructed cadets in drawing from 1833 to 1840. Eastman had used the opportunity to improve his own abilities as an artist, studying with well-known painters Robert Weir and C. R. Leslie, who also taught at

West Point. He even showed some of his paintings at New York City exhibitions, where his work was well received. During his West Point assignment, Eastman met and married Mary Henderson, the daughter of the academy's surgeon, Thomas Henderson.[8]

Mary Eastman accompanied her husband to Fort Snelling, and during their seven-year stay at Bdote, the couple developed a keen interest in Dakota culture: Seth Eastman as a painter, Mary as a writer. Her book, *Dahcotah, or, Life and Legends of the Sioux around Fort Snelling*, illustrated with his drawings and published in 1849, delivered to its readers the Eastmans' view of the Dakota.

Though neither of the Eastmans openly recognized the captain's previous marriage to Stands Sacred or his daughter Wakantankawin, there are hints, particularly in Mary Eastman's writings, that connections with her husband's

Dakota friends and relations still remained. In the introduction to her book, she explained that her sources for information were the "many friends" her husband had made during his first tour of duty at Fort Snelling. She was surprised to find that they were curious about her: "On going back to the Indian country, he met with a warm welcome from his old acquaintances, who were eager to shake hands with, 'Eastman's [wife].'" It seems quite possible that these were, in fact, Captain Eastman's relatives from his previous marriage, as well as other Mdewakanton from Cloud Man Village. According to Mary Eastman, they visited her family almost daily, and over time she learned enough Dakota to converse with them. With help from Philander Prescott, she was able to question her new friends about their customs and stories. Given all of this interaction, it seems possible that her husband's previous Dakota marriage was a known but unacknowledged family secret.[9]

Though Mary Eastman expressed great sympathy toward the Dakota, she found their behavior contradictory and puzzling. She was equally frustrated by the actions and attitudes of Americans, particularly the "frontier class of white people": "White men, Christian men, are driving them back; rooting out their very names from the face of the earth. Ah! these men can seek the country of the Sioux when money is to be gained: but how few care for the suffering of the Dahcotahs!"[10]

Still, she held as many of the beliefs in white supremacy and cultural superiority as her European American contemporaries. (Her second book, *Aunt Phillis's Cabin: or, Southern Life As It Is*, was published as a response to Harriet Beecher Stowe's *Uncle Tom's Cabin* and depicted slavery as a benign institution.) "They are receding rapidly," she wrote of the Dakota, "before the giant strides of civilization. The hunting grounds of a few savages will soon become the haunts of densely peopled, civilized settlements." Her main complaint about the behavior of other whites, beyond their

lack of compassion, was that they had not done enough to "civilize" the Dakota through education and religious conversion.[11]

Mary Eastman's description of her relationship with the Mdewakanton, condescending and ethnocentric though it is, nonetheless illustrates the frequency and complexity of interaction between the local Dakota and residents of the fort. For their part, the Dakota may have regarded the Eastmans and other officials at the fort as benefactors, despite or perhaps because of the dependence created by the 1837 treaty. The personal relationships that were so central to their culture and so important to past interactions with Europeans and Americans remained crucial to the Dakota. For most of the white community, however, these old relationships had changed. Though they may have held individuals in high regard or even affection, most whites saw the Dakota as a doomed race. Many believed it the "Manifest Destiny" of the United States to encompass all the lands of North America; they concluded that Dakota people would either be assimilated or become extinct. The Eastmans probably saw themselves as preserving the memory of a dying culture, a historical curiosity. In this they were mistaken. The Dakota had already begun to adapt, tentatively joining a market economy and trying out European American agriculture, and despite economic hardships, they held the main tenets of their culture closely, as even Mary Eastman noted.[12]

Captain Eastman enjoyed a productive time as an artist, creating seventy-five oil paintings and many more watercolors and sketches. Eastman's landscapes represent the best visual documentation of Bdote and its people from this era. Though he found himself in command of the post at various times, Captain Eastman probably had many opportunities to paint, for garrison duty during the 1840s was relatively quiet. When he arrived at Fort Snelling, only 129 officers and men were present—nearly one-third of whom were sick with fever acquired during the 1st Regiment's service in Florida. This left few men available for several major works of repair and rebuilding that were badly needed at the post. Twenty years of exposure to northern winters had taken their toll on many buildings, and the officers' quarters, originally a wooden structure, were to be augmented with a stone building.[13]

The work was further delayed by an incident on the Red River. In the summer of 1844, a party of drovers driving cattle from Missouri to Fort Snelling lost their way. Wandering into the Red River Valley, they blundered into a heated conflict between the Sisseton Dakota and the Métis, whose large-scale buffalo hunts had led to the quarrel. The Sisseton mistook the drovers for Métis, and in the ensuing confrontation, at least one of the Missourians was killed. In response to what was seen as an unprecedented attack on whites, a detachment of fifty men, the post surgeon, and several officers under the command of Lieutenant Colonel Henry Wilson—almost one-third of the garrison—was sent from Fort Snelling to Lake Traverse to bring in those responsible. They reached Joseph Brown's trading post at Big Stone Lake in October.

Though the Dakota were not overawed by the army's presence, they had no wish to see their trade interrupted by a war with the Americans. Brown, who was now the principal trader in the region, worked out an arrangement that allowed the Sisseton to turn over two warriors to the army in exchange for Lieutenant Colonel Wilson's agreement that the United States would send an expedition to drive the Métis out of Sisseton territory, thus allowing all involved to claim a degree of success. This was the largest expedition by troops from Fort Snelling since the Red Bird attacks in 1827 and illustrated again the limitations of the army operating away from its base. As in the past, negotiations by a third party were required to accomplish their mission. In any event, the prisoners escaped on the first moonless night during the march back to Fort Snelling.[14]

The long-delayed rebuilding project began in 1845 when "Lt. Granger with a party of 17 men left this place for the Chippewa country on the 1st of Feb for the purpose of cutting timber." The logs would later be sent in rafts to the government sawmill at St. Anthony Falls. The command of Fort Snelling at this time fell to Captain Electus Backus, who would oversee much of the rebuilding of both sets of barracks between April 1845 and October 1846. As the work got underway in April 1845, Backus reported to the adjutant general that the enlistments of a significant number of men at Fort Snelling were about to expire, creating "a remarkable deficiency of mechanics in the companies at this post," and that they would need to "employ citizens" or "send mechanics with the recruits designed to fill the vacancies about to occur." It was a complaint that might have come from the pen of Josiah Snelling twenty years earlier and retold a familiar pattern for frontier garrisons at this time. Every month would see the number of troops decline as men left the service, deserted, or died; a new batch of recruits would arrive, and the process would begin again. Fort Snelling's post returns plainly show this pattern: November 1844: 184 men present; January 1845: 179; March: 175; May: 161; July: 155; and by September 1845, after 28 men of Company G were transferred, only 88 men were on duty.[15]

In spite of the loss of men, Backus reported in August that "the stone work of the new Block of soldiers quarters is already completed. . . . It will be ready for the troops by the middle of October." Additional recruits did arrive the following spring, and by October 1846, Major Clark noted, "About the close of May last the troops at this Post commenced taking down the old block of quarters occupied by officers, and since that time have been much employed in procuring materials and aiding in the erection of a block of stone quarters for officers, the stone work of the walls is finished, the block under roof, one set of the quarters finished and occupied and the remainders of the building progressing towards completion as rapidly as the means at our disposals will permit."[16]

Some of the garrison might have wondered if anyone would be around to appreciate their handiwork, for events far removed from Fort Snelling and Bdote would soon have a major impact on the post.

⁓

In the spring of 1846, the United States went to war with Mexico, seeking to enforce its claim to Mexico's land north of the Rio Grande. Although many of the troops that fought for the United States were in volunteer regiments, there was also a strong need for regular army troops. Company after company from the 1st Regiment was sent from the frontier to the campaigns in Mexico. By March 1847, only Company D of the regiment remained at Fort Snelling. Captain Seth Eastman, as the only regular officer present, found himself in command of the post as well as acting adjutant and quartermaster. Thanks to the recent arrival of thirty-one new recruits, there were 101 soldiers on hand that month, but as in the past, this number steadily declined as men at the end of their enlistments were discharged and others deserted. By July, Eastman had become concerned that there were not enough soldiers to defend the post. Writing the commander of the 3rd Military Department, he requested that Company D of the 1st Regiment be reequipped with the new percussion cap muskets, replacing the old flintlocks: "The force at this post is weak, and it is desirable to render it as efficient as possible—it will be much more so with the percussion musket. . . . If the Col. commanding approves of this requisition, I recommend that the muskets now in the hands of the soldiers at this Post be placed in store for the purpose of arming the settlers near this place

in case there should be any trouble with the Indians on this frontier."[17]

While it was unlikely that any new weapons reached Eastman's troops before they were replaced the following year, replacement troops probably had the new muskets. Still, the letter reflects Eastman's concerns and the inability of the small garrison at Fort Snelling to intervene in the ongoing conflicts between the tribes.

And that potential for conflict had increased, thanks to the actions of fur trader and politician Henry Mower Rice. He had worked for the AFC at Mendota, assisting Franklin Steele, between 1839 and 1842, and then he became the company's trader to the Ho-Chunk, who had been moved from Wisconsin to Iowa Territory. As whites moved onto their new lands, the Ho-Chunk agreed in 1846 to move again. They trusted Rice and relied on his judgment, asking him to pick a good location for them. The trader assured the Ho-Chunk that the lands on the Upper Mississippi between the Crow Wing and Watab Rivers would meet all of their needs. This assessment was utterly untrue, as the new reservation was on swampy, heavily forested terrain unsuited to a people whose way of life depended on prairies. The primary needs being met were those of Henry Rice and the merchants of St. Paul, who welcomed the potential influx of annuity money from the new reservation, about a hundred miles up the Mississippi and far from the reach of other traders. It also sat, precariously, in the buffer zone between the Ojibwe and the Dakota.[18]

To oversee the Ho-Chunk reservation and prevent conflict with the Ojibwe, from whom the land had been purchased, the War Department decided to create a new post on the west side of the Mississippi six miles below the junction with the Crow Wing River.

In the fall of 1848, a party of "mechanics," workmen hired by the army, traveled by steamboat from St. Louis to Fort Snelling. From there

Fort Ripley, 1862. *MNHS.*

they traveled overland by wagon to the location chosen for the new post, a journey that took over nine days. Company A of the 6th Infantry, which had relieved the 1st Infantry at Fort Snelling in September 1848, acted as an escort for the workmen and also helped with construction. It was not a happy team. According to Charles Stearns, one of the contractors, "the company of Infantry after sojourning with us a month or more, (the weather getting verry [*sic*] cold) left for warmer quarters, returned to Fort Snelling, and left the Workman some fifteen, to take care of ourselves." The soldiers were no happier. Private Gustavus Otto, upon returning to Fort Snelling, wrote a friend, "What we had to endure there, cannot be described. I had got dysentery, from which I thought I would not recover. . . . [W]e marched back again . . . to . . . Fort Snelling, where we

remained. There it was the first time that I slept under a roof again, for until then we were treated like dogs." The new post, originally called Fort Gaines but soon renamed Fort Ripley, continued to be garrisoned and supplied by Fort Snelling for the next decade. Even though the Ho-Chunk would be moved again in 1851 (this time to Dakota lands south of the bend in the Minnesota River), the presence of troops at Fort Ripley provided a degree of stability and security important to the region's development. The colonists who flooded into the area incorporated the city of St. Cloud in 1856.[19]

The establishment of Fort Ripley also marked a transition in the role of Fort Snelling. Over the next decade, the fort would become a supply and support center, while its mission as a frontier outpost would diminish.

8. Endings

FORT SNELLING'S NEIGHBORHOOD WAS CHANG-
ing, and that change would only accelerate. Though
St. Paul had steadily attracted settlers throughout
the 1840s, initially growth was not explosive. Economic conditions, the war with Mexico, and the
overall perception of the region as unsuitable for
agriculture had tempered immigration. In 1850,
the city's population stood at 1,112, with an additional 226 living in the area nearby; a further 537
souls occupied the village of St. Anthony, near the
east side of the falls.

In the early 1850s, soldiers still visited old
Pig's Eye for drinking and social life, but it could
be a dangerous place. Even getting there might be
worth your life, as in the case of two soldiers heading to St. Paul "on a spree" who drowned when
their canoe overturned. More sinister things
might happen as well. In September 1851, James
Searles, a soldier recently discharged from Fort
Ripley with $800 in back pay, was "found dead in
the boom below the Lower Landing Saint Paul."
Though the coroner's inquest ruled his death a
drowning, there was no mention of what happened to his money.[1]

Not all was murder and mayhem, however.
The local merchants generally welcomed soldiers,
as they usually paid in cash. There was also a bud-
ding cultural connection between the garrison at
Fort Snelling and the people of St. Paul and other
nearby towns, in the form of the 6th Regiment's
band. Its leader and principal musician was Robert F. Jackson. By all accounts, Jackson was a virtu-
oso player of the keyed bugle and an accomplished
bandleader. Post returns and census records also
indicate that the 6th Regiment band was unusually large, with as many as nineteen musicians,
among the largest ensembles at any post in the
army. It can be little wondered that they made
a great impression on the citizens of Minnesota
Territory. In just over two years, they played in
several Fourth of July parades and numerous balls
and concerts around St. Paul and nearby towns.
James Goodhue, editor of the *Minnesota Pioneer*,
paid particular notice to the band's last concert,
performed from a steamboat as they left for a
new assignment at Jefferson Barracks on June 12,
1851: "Saint Paul sat upon the Bluff and listened
and dropped a silent tear as the last strain of the
Band faded away in the distance—a tear of fond
regret. Nothing so quickens tender recollections

The 1850 federal census for Minnesota Territory provides the first detailed picture of the population in and around Bdote in September of that year. Mendota continued to be an active fur trade community, dominated by French Canadians (42) and a considerable number of Minnesota-born women and children (87), most of them families of traders. Though the census did not indicate it, many if not all of them were probably of mixed European/Native ancestry. Their occupations—traders, voyageurs, blacksmiths, carpenters, and clerks—related to the fur trade. The one family showing American origins was that of American Fur Company agent Henry H. Sibley. This household included Sibley; his wife, Sarah, who was the sister of his business partner, Franklin Steele; her mother, Mrs. Mary H. Steele; and the Sibley children, along with Henry's brother Frederick. Also listed with the Sibley family was a forty-year-old Irish woman named only "Mary," probably a servant. Mendota remained a community as it had existed in the "Saint Peter's" region for fifty years or more.[i]

The community around Fort Snelling and within the garrison itself was dramatically different, however. With the exception of very young children and the family of interpreter Philander Prescott, nearly all were born in the United States or Europe. Americans—probably mostly attached to the Indian agency—made up the majority of those living and working on the reserve; many of them were from Ohio, Pennsylvania, and New York. Within the garrison of the post, 96 of 133 were foreign born. The Irish formed the majority, making up half of the immigrant soldiers.

There were at this time relatively few officers serving at the post, but they represented most of the small number of southerners at Fort Snelling.

Among them were Major Lewis A. Armisted and fellow Virginian Captain James Monroe; Lieutenant Richard Johnson was from Kentucky. Among the enlisted men only two, a private from South Carolina and another from Virginia, hailed from the South.[ii]

Living within Fort Snelling were seventeen families, wives and children of both officers and enlisted men. These included thirty children. The post must have felt like a small town at times: there were the thirty children in military families plus eleven children in nine other families living nearby. Indeed, the small schoolhouse on the parade ground must have been bursting at the seams, and the Reverend Gear, who served as both chaplain and schoolmaster, probably had his hands full.[iii]

A number of families clearly had free servants, although they were not listed as such. These appeared as single women living in the household, often of foreign birth, a number of them Irish. Similarly, some men attached to households were listed as "laborer." All of these individuals seem to have been whites of European or American origin. There was at least one enslaved worker and possibly others still living and working at Fort Snelling in 1850, but slaves were listed separately on the 1850 census, and there were no slave schedules for Minnesota Territory. In pay records the following year, however, Major Armisted listed an enslaved servant, Sarah Ann, at Fort Ridgely, and it is likely she was present at Fort Snelling in 1850.[iv]

This picture of the population around Bdote in 1850 is, of course, incomplete, since it did not count the considerable numbers of Dakota living in nearby villages, including those of Penichon, Black Dog, and Shakopee within the boundaries of the military reserve.

and fond remembrances, as the music to which we have often listened; its last notes are like sunset to the heart."[2]

But the fort and its garrison would soon confront a booming American and European population on its doorstep.

Almost immediately after the creation of Minnesota Territory in 1849, its political leaders began working for a new treaty to acquire the remaining Dakota lands. The primary drivers of this effort were Territorial Representative Henry Sibley and newly appointed Territorial Governor Alexander Ramsey. Though representing competing political parties—Sibley was a Democrat and Ramsey a

Whig—the two men understood that the path to statehood depended on taking the land from the Dakota. Sibley used his influence with both the traders and the Dakota, seeking to win them over to the idea, while Ramsey used his connections with the administration of President Millard Fillmore to gain support in Washington.[3]

In the summer of 1851, the US government's representatives negotiated two treaties with the Dakota. The first was signed at Traverse des Sioux in July with the Sisseton and Wahpeton bands, and the second at Mendota with the Mdewakanton and Wahpekute in August; the fort's garrison provided a small escort at Traverse des Sioux. Like most treaties negotiated in the mid-nineteenth century, the proposal was presented to the Dakota as being in their best interest and necessary for their survival, yet key provisions managed to mostly benefit the interests of traders, colonists, and the government. The Dakota gave up 35 million acres of bountiful lands that had supported their people for centuries, retaining reservations on the Minnesota River—ten-mile-wide strips of land running for 140 miles on each side of the river. They were promised cash payments, annuities, and other considerations, similar to the 1837 agreement, including $275,000 to cover costs of their removal and the settling of fur trade debts. Dakota signers marked two copies of the treaty, then were led to a third table, where they

St. Paul, 1851. The First Baptist Church, located at what was then known as Baptist Hill (now Mears Park), is in the background; cordwood is stacked by the fence in the foreground. *MNHS*.

Artist Francis Blackwell Mayer sketched many scenes at the signing of the Treaty of Traverse des Sioux in 1851 and later made this oil painting. *MNHS.*

signed a document giving that sum directly to the traders, without any discussion of the fairness of the traders' claimed debts. Many European Americans who were present, like the Dakota signers, also had no understanding of this document, the "traders' paper."

The Dakota agreed to the provisions only after long, sometimes rancorous negotiations and unrelenting pressure from the US government's commissioners. Many Dakota leaders pointed out that the government had never fulfilled its 1837 treaty obligations, and they did not believe its representatives would do any better this time. Goodhue recorded Chief Wabasha's declaration that vague promises about schools and other amenities were no substitute for cash: "In the treaty I have heard read, you have mentioned farmers and schools, physicians, traders, and half-breeds. To all these I am opposed. You see these chiefs sitting around.

They and others who are dead went to Washington and made a treaty [in 1837], in which the same things were said; but we have not been benefited by them, and I want them struck out of this one—we want nothing but cash turned over to us for our lands."[4]

Expectations among the citizens of St. Paul soared that summer. "The news of the Treaty exhilarates our town," exulted Goodhue in the *Minnesota Pioneer.* "It is the pillar of fire that lights a broad Canaan of fertile lands. We behold clearly, in no remote perspective, like an exhibition of dissolving views, the red savages, with their teepees, their horses, and their famished dogs, fading, vanishing, dissolving away; and in their place, a thousand farms, with their fences and white cottages and waving wheat fields . . . and villages and cities crowned with spires, and railroads with trains of cars."[5]

The new treaties rapidly changed Fort Snelling's situation. Growth on the newly ceded lands would mean more growth in the area around the confluence. As speculators eyed the fort and its reserve lands—plums in the expanding real estate market around St. Paul, St. Anthony, and Minneapolis—its very existence would be questioned. And the fort's relations with local Americans, particularly newcomers, were not always smooth. While Goodhue had a soft spot for regimental bands, he could be scathing toward military inefficiencies. He disdainfully noted that soldiers were farming to support the garrison: "Is not the drill agricultural, a new feature in military tactics?" And when a group of Dakota being escorted from Benton County in chains to stand trial for killing an Ojibwe escaped, Goodhue was livid: "What are officers and men kept dressed up in regimentals for, at our garrisons, at an immense expense? The Government had better disband the troops, and recruit the service with a regiment or two of women, who have had the courage to put on the Turkish pantaloons and short skirts."[6]

The immediate impact of the treaties was negligible. Until the treaties were ratified and the lands surveyed, the west side of the Mississippi was closed to settlement. Ratification was slow, with resistance coming once more from southern senators leery of expanding a territory that might become another free state and upset the balance of power in Congress. The Dakota, who were not expected to move until after ratification, still occupied their villages, hunting on their traditional grounds, and even newspapers commented on their continued visits to St. Paul.[7]

For more than two years following the treaties of Traverse des Sioux and Mendota, an unsettled period of transition pervaded the region. Each part of the community—Dakota people, the newcomers eager to take the land, and the military

The Sioux Indians of Kaposia, being solicitous [concerned] about the ratification of the Treaties, assembled last Saturday in large numbers . . . and had a sort of informal council. The Chief Little Crow . . . stood up and talked as follows:

Last year, our Great Father made a great fuss about the treaties. He asked us to hitch along and let him sit down on our grounds, while we could have a talk together, about his buying our land. Our Father sat down with us and began to talk with us and whittle a stick and then whistle and he kept on in that way for almost two moons. . . . We got very tired. We danced attendance on our Father, so long that we raised no corn. Our Father is a devil of an old fellow to hunt, if he can only corner a drove of cattle. Our Father is a great glutton; he would go and shoot a cow or an ox every morning and give us the choice pieces of it, such as the head and the paunch; and there he kept us waiting for six weeks; and when the cattle were nearly all gone, and he had whittled all the sticks he could find, he got up and shut his jack-knife, and belched up some wind from his great belly, and poked his treaty at us, saying, I will give you so much for your land. It is true he said that the Senate at Washington, would have to ratify the treaties. Well, we signed the treaties. We could not help ourselves. We went home. We had no corn crops and could find no game to speak of as well, the White settlers came in and showered down their houses all over our country. We did not really know, whether this country any longer belonged to us or not. . . . But this is what we are waiting to know, whether our Father means to take our lands for nothing or whether he means to pay us the money and the annuities he promised us in the treaties? We do not want to be humbugged out of our lands. We owe debts and we want to pay them. If our Father had said, "move along; you must move along; you shall move along." It is likely we should have had to go, but that was not the way our Father talked to us. He said, "You have no game here, our people are hemming you in, you can have no schools for farming while you live scattered, you owe debts, you need annuities; will you go, my Red Children, if we give you so much?" We thought that was very kind and we said yes. Now what have we? Why, we have neither our lands where our fathers' bones are bleaching, nor have we anything. What shall we do?[v]

garrison—had its own radically different perspective on the future.

For the Dakota, this was a time of enormous uncertainty and palpable loss. The government's long delay in ratifying the treaties and furnishing the promised annuities was troublingly reminiscent of what happened after the 1837 treaties—as were the incursions onto their land by whites who were not waiting. One of the few reports of their feelings comes from an account of a speech given by Little Crow, grandson of the man who signed the agreement with Pike, son of Wakiŋyaŋ Taŋka, and now the leader of Kaposia Village (see sidebar page 123). Given the difficulties of translation, of course, this speech should be taken with a cautious grain of salt, but the words have an authentic character to them—and the speech accurately outlines the predicament of the Dakota and reflects their feelings about the treaties.

Southern politicians, continually focused on the balance of power in Congress, did more than merely delay the treaties. After the Senate finally ratified the documents in June 1852, the Dakota would learn of amendments that were intended to make the treaties so unacceptable that the chiefs would refuse to sign the final versions. Key provisions, such as the creation of a permanent reservation and the size of the cash payments, were changed. The reservations became vague, temporary arrangements "at the pleasure of the President," and the cash payments were greatly reduced. The Dakota leaders resisted the changes. Steven Riggs recorded Wabasha's bitter summary of the reaction: "There is one more thing the great father can do, that is, gather us all together on the prairie and surround us with soldiers and shoot us down." But in the end, the Dakota had no choice but to accept a deeply flawed document that deprived them of their homeland. This cynical and corrupt agreement, fueled by lies and deception, was forced on the chiefs against their better judgment and desires.[8]

Ta Oyate Duta or Little Crow, 1862. *Photograph by Whitney's Gallery, MNHS.*

Some traders, like Hercules Dousman, were embarrassed by the blatant double-dealing surrounding the treaty; he noted somewhat prophetically, "The Sioux treaty will hang like a curse over our heads the balance of our lives." Yet barely a ripple of public concern seems to have troubled the boosters in St. Paul. In the autumn of 1853, when the Dakota eventually moved to the reservations specified by the federal government, Goodhue of the *Minnesota Pioneer* expressed some regret for the missing Dakota friends seen around town and given belittling nicknames like "One Legged Jim" and "Old Bets" (Azayamaŋkawiŋ, or Betsy St. Clair) who had seemingly "passed away," with the rest of their tribe. But his feelings for the Dakota as a people lacked any remorse:

This is the consummation of another link in the great chain of their destiny. The wants of the whites have demanded this sacrifice, as they will require another a few years hence, to be again followed by another and another, until the race becomes extinct, and the vast country west of us becomes populated by a more useful class. The land which now provides for the Indians a precarious subsistence . . . will be brought under the control of the agriculturalist, the miner and the mechanic . . . furnishing the luxuries as well as the necessaries of life, for many portions of the world.[9]

The whites would indeed continue to require "sacrifices" from the Dakota. But the Dakota would not go quietly, they would not become extinct, and they would continue to live on their ancestral lands.

Whites, particularly immigrants from the East, were anxious to colonize the ceded lands, but they knew that acting before the treaties were ratified and the lands surveyed was a gamble. Preemption claims—unsurveyed land taken and improved with the expectation of a later purchase—could be lost, and the Dakota might be hostile to such premature ventures. Nevertheless, for many whites, the drive to secure the best lands and the most advantageous locations prevailed, and homesteads quickly began appearing on the west side

A Dakota family camped at what is now Marshall Avenue, St. Paul, in about 1860. *MNHS.*

of the Mississippi and along the valley of the St. Peter's, officially renamed the Minnesota River in 1852.[10]

With the opening of the former Dakota lands, St. Paul was becoming a boom town. Thousands of land-hungry immigrants, not only from eastern states but increasingly from Europe, were crowding onto steamboats bound for Minnesota Territory. Towns began appearing in the Minnesota River Valley, extending ever further into what had been Dakota lands only a few years earlier. As treaty lands were surveyed by the US government and land offices opened, tracts began to pass into private ownership, and as population from immigration increased, the land prices rose. Properties purchased from the government at the standard rate of $1.25 per acre one year might sell again for $5 per acre the next. Speculators, some buying directly from the land office, others cashing in bounty land warrants, were making spectacular profits. By the end of 1855, a land boom was underway. Historian J. Fletcher Williams, who arrived in St. Paul in 1855, was a witness. "The fever of real estate speculation . . . began to grow into a mania which a few months later almost rendered St. Paul a by-word." Year by year this fever grew; though most of the early land dealers were reasonably honest, by 1857, the land boom had attracted a class of reckless traders and "confidence men," "having no office but the sidewalk and no capital but a roll of town-site maps and a package of blank deeds, yet all fairly coining money, and spending it, in many cases, as rapidly as made on fast horses, fast women, wine and cards." This boom came in the midst of a rapid increase in population. From 1850 to 1860, the total number of residents (not counting Native Americans) swelled from 6,077 to 172,023.[11]

Fort Snelling and increasingly the military reserve lands around it were regarded less as a source of security and stability and more as an impediment to local development. Following

the 1837 treaty and Plympton's clearing of the military reservation, squatters began steadily encroaching on the reserve lands on either side of the Mississippi. By the 1850s, a variety of buildings, from claim shanties to substantial homes, could be found throughout the area. There were so many conflicting claims that the squatters organized claim associations to defend themselves and lobby for a change in the reserve boundaries. Accordingly, a bill to reduce the size of the Fort Snelling reservation by nearly two-thirds was introduced in Congress in 1852 and passed that August. The new law shifted the reserve's northern boundary several miles to the south, from Nine Mile (Bassett) Creek above the Falls of St. Anthony to Browns Creek, today known as Minnehaha Creek—several miles below the falls and considerably closer to Fort Snelling. While Pike Island remained in the reserve, the lands on the east side of the Mississippi, with the exception of a small strip at the ferry landing opposite the fort itself, passed into public domain, as did the government mills and strategic waterpower sites bordering the Falls of St. Anthony.[12]

Among the chief sponsors and supporters of this new act were Henry Sibley and Henry Rice, who both had personal and civic interests in its passage. In March 1853, a Military Reserve Claim Association was organized in St. Paul to protect the claims of those who had preempted lands on the former reserve. It was one of several claim associations on the military reservation; such organizations were common in the Midwest and were intended to keep speculators or "claim jumpers" from purchasing lands out from under those who had established preemption claims. But this particular claim association seems to have been formed for the very purpose of speculation, since its members included Henry Rice, John Irvine, and William R. Marshall, all of whom were heavily involved in the frenzy of the land speculation then overrunning Minnesota Territory.[13]

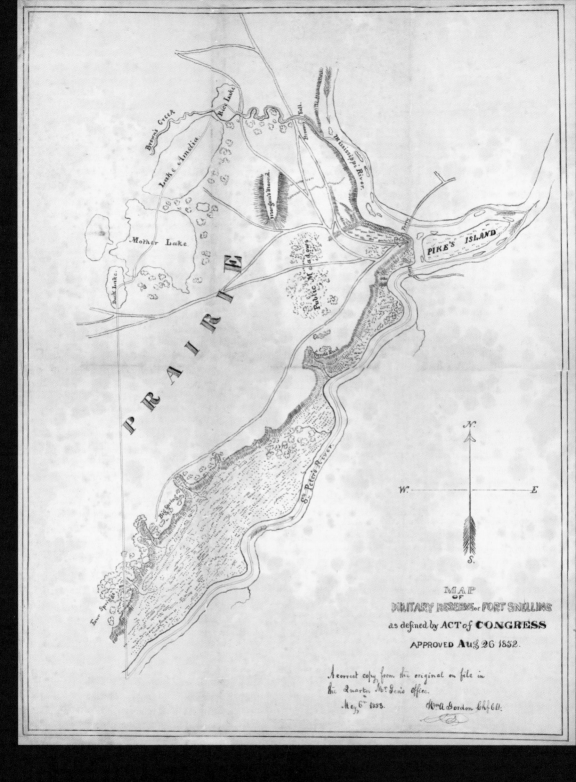

A map of the reduced Fort Snelling military reserve, 1852, copied from the original on file in the quartermaster general's office, May 6, 1853, by William A. Gordon, chief clerk. *MNHS.*

White frustration with the continued presence of Indigenous people reached a high point in St. Paul in 1853 with an incident in a back-and-forth series of retributive attacks between Ojibwe and Dakota bands. A number of Ojibwe men "took their stand in front of [Forbes's] store and fired in through the open door, badly wounding one of the [Dakota] women....A more daring feat was [n]ever perpetuated by any people than this attack of the Chippewas, in the city of St. Paul, in the most business part of town and in broad day light." The governor ordered Company D of the 1st Dragoons, recently stationed at Fort Snelling, to track down the offending Ojibwe, but for editor Goodhue's taste, the troops were "a little tardy"; he noted approvingly that an impatient body of citizens "under the command of Lt. Simson, started after the Indians."[vi]

Simpson, a captain in the topographical engineers who happened to be in St. Paul on other duties at the time of the incident, wrote the paper a few days later, setting the record straight: "[T]he motive which prompted me to go in pursuit of the Indians, was the very natural one of desiring to apprehend the offenders.... In my judgment the dragoons, considering the distance they were from the scene of outrage, the difficulties they had to encounter in crossing the river, and the long chase they intended to give the Indians, were in the saddle as promptly as could have been expected." Simpson concluded with a subtle rebuke: "Knowing the feelings of my brother officers as I do, and how important it is that they as well as every one else, should be rather encouraged to generous deeds, by liberal remarks, than be depressed by those which may bear a contrary construction, I have felt called upon to express myself as I have."[vii]

The changing of the reserve boundary also opened up development around St. Anthony Falls, ending what had been seen as a major roadblock to the region's economic progress. Since 1849, a group headed by Congressmen Robert Smith of Illinois and Cadwallader Washburn of Maine had leased the government-owned sawmill and gristmill near the falls from the War Department.

They and a small number of investors, mostly Washburn family members, were well positioned to begin rapid development of the waterpower potential of the site. By 1855, the village of Minneapolis began to thrive on the west side of the falls.

⏤

During the 1850s, the troops at Fort Snelling continued to assist the work of government officials and Indian agents by escorting treaty negotiators and being a visible presence at such occasions. They provided similar support at annuity distributions. With the removal of the Dakota to the Minnesota River reservations, the government began construction of a new post at "Little Rock"—Fort Ridgely—to keep the peace as colonists poured into the area.

Companies C and K of the 6th Infantry were transferred from Fort Snelling to help with construction. To replace them, Battery E of the 3rd Artillery was assigned to the fort. Known as the "light battery," Battery E operated with horse-drawn guns or "field artillery," as opposed to large guns in fixed emplacements found in most coastal fortifications. Gun sheds to protect their cannons from the elements were subsequently constructed at Fort Snelling.

In May 1854, Company D of the 1st Dragoons was reassigned to Fort Leavenworth, Kansas Territory, after five and a half years' duty at Fort Snelling. Soon after their departure, on the night of May 27, a drunken soldier set fire to the dragoon stables at the post, destroying the building along with five wagons, two carts, several sleds, and a ton of stored hay. The soldier was presumed killed in the blaze, as a man was reported missing the next morning. During the next year, the size of the garrison reached a low point.[14]

With a small garrison and a growing local civilian population, the army began to shift away from relying exclusively on the labor of troops to support the post. The quartermaster began to hire laborers, mechanics, and carpenters for repair and construction at Fort Snelling as early as 1850. In 1854, he contracted with Franklin Steele for four hundred cords of wood. In his return for August 1854, Captain Thomas Sherman, commander of the light battery and of the post, reported, "To keep a battery at a post like this with but 45 men and attend to all necessary duties required of the company, is next to an impossibility, more recruits are respectfully solicited." The following year, wood, hay, grain, and charcoal were purchased from St. Paul suppliers. There were simply too few men to maintain the military integrity of the fort

and continue duties like woodcutting, haymaking, and gardening, as they had in the past.[15]

When Companies A, D, I, and K of the 10th Infantry, under the command of Colonel Edmund Alexander, joined the garrison in October 1855, Fort Snelling could hardly be described as a "frontier" outpost. To many observers, it must have seemed that the original reasons for the fort's establishment thirty-five years earlier no longer applied. America's border with British Canada had long since been determined and disputes resolved. The fur trade had largely evolved into the "Indian trade"—the practice of advancing goods to Dakota

Fort Snelling interior, as painted in 1853 by George P. Fuller, a civil engineer assisting on a topographic survey of the area around the fort. The buildings are, from left, the hospital, school, magazine (with the round tower behind it), and well. *Watercolor, MNHS; gift of H. W. Morris.*

and Ojibwe customers, then collecting from annuity payments for the debts—which did not seem to require protection. Many of the traders, like Sibley and Dousman, were moving into other ventures, including real estate, lumber, and transportation. After the 1851 treaty, most of the actual trade in furs was taken over by firms like Ullmann and Rose, who purchased pelts directly from settlers and white trappers for sale in the East.[16]

By 1857, civilian employees began to appear on post returns as part of the quartermaster department. That January, these included a clerk, a forage master, a blacksmith, and twenty-seven teamsters. The presence of so many teamsters and a forage master to procure hay and grain indicates to what extent Fort Snelling was developing into a supply and distribution center for Forts Ripley and Ridgely, relying on local civilian labor.[17]

American residents of St. Paul and the other towns and villages near Bdote continued to question why the fort, even in its much smaller state, still took up valuable real estate in their midst. If there had been a threat of an Indian "outbreak," surely that had now diminished. Indeed, many in the settler community believed that a new day had dawned since the 1851 treaties were ratified, and the Dakota would "not many years hence be found on the other side of the Rocky Mountains."[18]

In reality, immigrants into the new US territory found a more complex situation. Far from vanishing into the sunset, Dakota and Ojibwe people continued to traverse their homelands to hunt, fish, and gather resources, even if their villages had moved. The Ojibwe had specifically reserved these rights in their treaties; the Dakota simply sought survival. Their presence caused a certain amount of consternation among European and American colonists, who had expected to find an "untamed wilderness" devoid of human occupants. Their assumptions arose in part from the press accounts of boosters, which omitted any reference to the Natives, and the centuries-old European American concept that land not under agricultural use or some other exploitation was being wasted, and was thus wilderness. Sometimes interaction with Indigenous people alarmed new immigrants unfamiliar with Native culture. Language barriers, different concepts of etiquette, and different social customs led to misunderstandings. In both Ojibwe and Dakota societies, for example, it was not unusual for visitors to enter another family's dwelling without invitation or announcing their presence in advance; casual conversation was not expected of guest or host. White settlers found this behavior rude and threatening. For Dakota and Ojibwe people, sharing food was an important cultural bond, and refusing to share was considered very rude—even uncivilized. Whites saw the request or expectation of food as "begging" and demeaning conduct. In the words of historian Bruce White, to the Indians, settlers' reluctance to share seemed "cold and ungenerous. . . . By native standards whites were relatively prosperous people with goods and food stored up in log houses, who shared grudgingly if at all."[19]

In a harsh climate where people often had to rely on neighbors to survive, understanding and even friendships sometimes developed between Natives and whites, but these relationships were not the norm. Friction and anxiety, added to the racial attitudes held by the great majority of whites, created the potential for conflict.

For soldiers at Fort Snelling, the winter of 1855–56 was somewhat unusual. From October through May, the post held from 370 to 380 troops, the largest garrison since the 1820s. However, when summer arrived and transportation by road and river improved, the 10th Infantry began dispersing its companies to garrison other posts. Companies C, D, K, and I went to Fort Ridgely, Company A to Fort Ripley, and Companies B and G to Fort Crawford. This left only the headquarters, band, and Company H of the 10th Infantry and Battery E at Fort Snelling, about 130 troops in all. The proximity of St. Paul and other towns probably encouraged a spike in desertions, with eighty men leaving between March and July.[20]

The reduced garrison sometimes strained to carry out its mission. In June 1857, Territorial Governor Samuel Medary and Indian Agent Charles Flandreau requested Captain Thomas Sherman, then in command at Fort Snelling, to furnish "fifty mounted men" to accompany officials to the distribution of annuity payments on the Dakota reservation in early July. Captain Sherman had no mounted men to supply, since the company of dragoons who had accompanied officials on previous occasions had been transferred three years earlier. "We have only a company of artillery," he wrote. "None of it is mounted and equipped as dragoons, nor have I the means of so mounting it." He proposed instead to take the battery with "a guard of infantry troops drawn from Fort Ridgely" as escort.[21]

This request for an escort was more than a casual formality. The previous winter had seen unprecedented violence on the southwestern border of Minnesota and Iowa, an explosion that followed years of trouble. In 1853, a fur trader named Henry Lott had murdered a Wahpekute leader named Red All Over (Sintomniduta) and nine women and children in his family. When his brother, Scarlet Point, better known as Iŋkpaduta, sought justice from civil authorities at Fort Dodge, the prosecutor scoffed at the idea of bringing charges and nailed Red All Over's head to a post outside his house. When Iŋkpaduta then approached the military at Fort Ridgely, a single patrol failed to find Lott, and the matter was dropped. In March 1857, after conflicts over food and resources during a particularly harsh winter, nervous whites forced Iŋkpaduta's band to turn over all its guns—tools that were necessary for survival. This band then recovered the weapons and began a series of raids. Iŋkpaduta and his followers attacked homesteads and trading posts from Spirit Lake, Iowa, to Springfield (now Jackson), Minnesota, killing forty white immigrants before heading west beyond the borders of the territory.[22]

For the first time in more than fifty years of interaction between Americans and Dakota, white settlers had been killed. Previous incidents had involved disputes between individuals, and not random attacks. Troops from Fort Ridgely gamely tried to pursue the killers, but the ill-equipped company of infantry had no hope of catching them.

These events, which came to be known as the Spirit Lake Massacre, provoked unrest bordering on panic within the white residents of Minnesota Territory—and troubled the Dakota, as well. By the mid-nineteenth century, the people of the Seven Council Fires inhabited a vast region from the Mississippi to the Yellowstone Rivers, but they experienced increasing pressure from a steadily expanding United States. While the Dakota of Minnesota were parting with their ancestral lands, their western relatives, the Lakota, were among the tribes agreeing to the Fort Laramie Treaty of 1851, which defined boundaries among western tribes and permitted white immigrants to pass through Lakota lands. Though this was intended to prevent trouble, the presence of colonists and soldiers eventually led to conflicts resulting in what the Americans called the First Sioux War in 1855-56.[23]

At the same time, the Sisseton and Mdewakanton in Minnesota experienced divisions that reflected where they lived: some in the west, like Iŋkpaduta and his followers, sought to live a traditional life as much as possible, and others living closer to Bdote, including prominent leaders like Little Crow, sought some accommodation with whites by taking up farming and adopting European dress. These divisions were worsened by the policies of the Indian agent, Charles Flandrau, whose actions favored farmer Indians. So in this atmosphere, fearing retaliation and goaded by government threats to delay or stop badly needed annuity payments, Dakota leaders agreed to send out a party to track down Iŋkpaduta. Though several of Iŋkpaduta's warriors were reportedly killed, all attempts by the eastern Dakota or the US military to capture him and his followers failed, and they escaped to live among the Yankton in Dakota Territory.[24]

The entire incident only increased tensions. While many Dakota disapproved of what Iŋkpaduta and his band had done, others resented being pressured to track down their own kin—and to some extent sympathized with their plight. White officials and settlers, ignoring efforts by the Dakota to track down Iŋkpaduta, were generally horrified by what had occurred, as it seemed to confirm their worst nightmares and stereotypes. They were also distressed by the army's inability to prevent the attacks or capture the perpetrators.

⸺

Regardless of tensions on the frontier, Minnesota political and business figures continued to push for a further reduction of the military reservation or the complete removal of Fort Snelling. This in itself was not an unusual or unreasonable expectation. Other posts built on what was originally the borderland were eventually surrounded by settlements, decommissioned, and sold. A notable example was Fort Dearborn, an important outpost built in 1803 on the Chicago River at Lake Michigan. By 1837, this fort was surrounded by the burgeoning city of Chicago; it was decommissioned and sold two years later. Land speculators both in Minnesota Territory and outside it expected Fort Snelling to meet a similar end. And with the acquisition of the Dakota lands and rapid population growth around St. Paul and the Falls of St. Anthony, they anticipated that end would come soon.

Franklin Steele, who had been speculating in Minnesota lands for nearly twenty years, was foremost among those expressing an interest in the fort and the remaining reserve lands. With the help of a key political ally, he sought to nudge the government to make a quick decision in his favor. In April 1856, when land speculation in Minnesota was white-hot, Territorial Delegate Henry Rice wrote Secretary of War Jefferson Davis asking the War Department to consider "disposing of the military reservation at Fort Snelling," enclosing in his request a proposal to purchase the remaining five thousand acres of the reserve for $15 per acre, a total of $75,000. Davis forwarded the plan to the Army Quartermaster General Thomas Jesup to provide the army's view on the continued value of Fort Snelling. General Jesup admitted that the fort was "no longer of any value as a position of defense" but noted that it remained important as a depot for gathering and transporting supplies for frontier posts. He stated his opinion that the War Department should decline the offer, and there the matter remained for a time.[25]

But 1856 was an election year, and in March 1857, a new president, James Buchanan, came into office. Buchanan's secretary of war was John B. Floyd, the former governor of Virginia. Steele and Rice decided to renew their proposal with a slight alteration, leaving the fort and forty acres surrounding it out of the purchase. Rice wrote the War Department once more, in April 1857, with Steele's new proposition. Again the proposal went from the War Department to Quartermaster General Jesup. Jesup once more demurred, arguing that Fort Snelling remained valuable as a point of concentration where troops could have winter quarters at a location more easily supplied than remote, frontier garrisons. And, in fact, this seems to have been the practice, as post returns from 1854 to 1857 show a pattern of companies moving to Fort Snelling in autumn and back to Forts Ripley and Ridgely in spring and early summer. But Secretary Floyd clearly disagreed with the army's position. Commenting on a later report by officers, including Commander of the Army Winfield Scott, who opposed the sale of Fort Snelling, particularly in light of the Spirit Lake Massacre, Floyd wrote, "Whatever may be the opinion of a 'military man' as to keeping a fort in the heart of a settled country to hold Indians in subjection who roam through the forest many days journey away from it, a man of common apprehension would conclude that the proper place for forts and troops was amongst the Indians. . . . When the United States army are set to 'cutting and stacking hay' it is possible a still more eligible point may be selected for the purpose near to the Indian settlements."[26]

Floyd went even further, not only overriding the army's concerns but sending Major Seth Eastman for his third assignment to Fort Snelling under confidential verbal orders to survey the reserve into various lots in preparation for sale. The major had hardly begun his work when an agent hired by Secretary Floyd, William Kent Heiskell, arrived in St. Paul to negotiate the sale of the reserve and the fort. Acting quickly and in total secrecy, he arranged to sell the property to Franklin Steele for $90,000, with one-third payable in July and the remainder to be paid in two equal annual payments thereafter. With Heiskell and Eastman signing as commissioners for the government, the sales contract was completed on June 6, 1857.[27]

Even by mid-nineteenth-century standards, this deal reeked of insider collusion. When it became known, a few days after the contract was signed, reaction from the public was vehement, particularly from those who also had an interest in the Fort Snelling lands. The editor of the *Chatfield Republican* had strong thoughts:

> The facts connected with this Fort Snelling Reservation sale are these. Eight thousand acres of land of the best quality, lying between two of the most important cities of Minnesota . . . and only ten miles asunder, are sold at private sale for the paltry sum of eleven dollars per acre, when it is a fact well known to every body acquainted with the land and its locality, that it would have brought readily at public sale at least $50 per acre. Take this, in connection with the fact that different citizens of the US had petitioned the war department to give public notice when the sale of said land would take place . . . and the transaction assumes at once its proper character as an atrocious swindle.
>
> . . . While we have a high opinion for Franklin Steele, we do not believe one of his virtues, is that of buying government land merely to benefit the settler, who may come along and desire to occupy it. As a speculation, and for speculative purposes only, did he buy it, and it is the silliest twaddle that ever [a] hired pimp put forth to put any other construction upon it.[28]

While many could complain, there was at least one of those "different citizens" with an interest in the Fort Snelling reserve who could do something about it. Congressman Robert Smith of Illinois, who had held the concession on the government mills at St. Anthony Falls since 1849, had a keen interest in the lands of the Fort Snelling reserve. He had already been in contact with Secretary Floyd about their probable sale, and it is likely he knew of Steele's earlier offer. Smith demanded a congressional investigation into the sale, and he had the clout to make it happen. Approval of the sale was delayed while Congress held a long series of hearings between August 1857 and May 1858. The result was a final report, 456 pages long, revealing a complex arrangement between multiple parties coming together under mysteriously convenient circumstances to arrange the sale.

Franklin Steele's silent partners were three individuals who made up the "New York Company": John C. Mather and Richard Schell, both New York politicians, and Dr. Archibald Graham, a Virginian with an interest in western investments. In April 1857, about the time that Steele and Rice were presenting their second offer to Secretary Floyd, Dr. Graham was meeting with Floyd, a fellow Virginian, about lands in Minnesota,

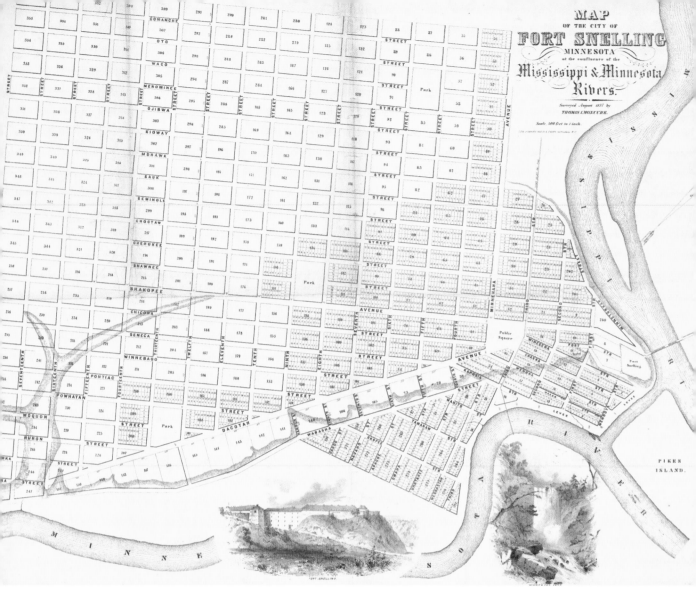

specifically the military reserves at Fort Snelling and Fort Ripley. In one of the happenstances surrounding the sale, Graham brought his New York partners into an agreement with Steele, prior to his final negotiations with Heiskell. The parties agreed not to pay more than $120,000 for the reserve. Franklin Steele would receive one-third of the property, Graham one-ninth, and Mather and Schell would divide the rest. In July 1857, when the sale was approved by the War Department, a first payment of $30,000 was deposited in the US Treasury—$10,000 from Steele and the balance from his partners. Heiskell's role in the matter also came under some scrutiny, as he too

was a Virginian and a political ally of former governor Floyd.[29]

Some sort of prior arrangement among Floyd, Heiskell, Graham, and the New Yorkers with Steele and probably Rice seems likely, but the real focus of the congressional investigation centered on the price the government received for the lands. Here testimony varied widely and seemed to depend on whether those testifying had a prior interest in the reserve lands. Those who were left out of the bargain tended to testify that they would have paid much more; others more removed or friendly toward Steele felt the price was fair. In the end, Congress produced two sets of recommendations: a majority view that the sale should be negated and a new, open auction held, and a minority report recommending that the sale be approved according to the original agreement with Franklin Steele. This choice might have seemed straightforward, with the majority view prevailing, but numerous petitions from all over Minnesota Territory had arrived in Congress urging that the sale be approved. After long debate, the Congress made no decision. The matter was referred back to Secretary of War Floyd.

Seeking some further justification for his decision to make the sale, the secretary ordered a board of officers to meet at Fort Snelling to determine if the army should continue to maintain the fort as a depot for supplying frontier posts and maintain a garrison there "to repel Indian incursions." The board included a mixture of officers from the cavalry, infantry, quartermaster department, and topographical engineers. For three days beginning on April 28, 1858, they gathered information on the issues to be examined, though their sources were not mentioned in the board's brief report. In the end, the board determined that Forts Ripley and Ridgely were "adequate . . . to satisfy all the requirements to be made at this time for the safety of the frontier; and it is not probable that any large increase in the number of posts or troops will become necessary at any Citizens' Military Training future period." Clearly, they did not see the Spirit Lake Massacre, which had occurred a year earlier, as an indication of potential problems in the future. Further, they noted that St. Paul, as "the head of ordinary navigation upon the Mississippi, would be the most suitable point of deposit and transhipment [*sic*] for the posts on the frontier to the north and west," avoiding the extra time and expense of unloading supplies at St. Paul and transporting them to the fort for storage and later shipment to Ridgely and Ripley. In conclusion, "The board respectfully suggest the entire abandonment of Fort Snelling, both as a military station and depot, and the establishment of an agency in St. Paul to forward the necessary supplies to the posts on the frontier by means of private transportation."[30]

This conclusion was all Secretary Floyd could have hoped for. The sale of Fort Snelling to Steele was approved, and on June 1, 1858, the last garrison lowered the flag and left "old" Fort Snelling. The post's thirty-nine years of service seemed over. On July 19, the post quartermaster transferred ownership of the fort and its reserve to Franklin Steele, who had already, with his partners, made surveys and laid out plats for the town of Fort Snelling. But by then, their plans proved to be a mirage.[31]

In August 1857, the nation had been rocked by a severe financial panic. The Minnesota land price bubble, inflated by six years of exuberant speculation, was well and truly burst. The crash ruined many of the local banks along with citizens who had invested heavily in property, including many "old settlers." By the summer of 1858, with no cash or loans to be had, there were few buyers for Mr. Steele's new town. Out of the ruins of his plans, Steele did manage to build a nice farmhouse just outside the walls of the former post and was said to have pastured his sheep on the old parade ground. It had been abandoned without much fanfare, only lamented, if at all, by the army. The quartermaster captain who oversaw the final removal of supplies and sale of surplus items wrote in his last report to General Jesup, "Thus terminates my mission, and thus terminates the military character of old Fort Snelling. . . . No more shall the flag wave over its venerable walls—no more shall the thunder of cannon declare the power that erected it—no more shall martial music quicken the step of the soldier within its hallowed precincts. It is gone. The reflection is a melancholy one—when we see the 'tooth of speculation' fasten upon the vitality of such a cherished spot."[32]

Snelling from left bank Miss river 1864

Fort Snelling from across the Mississippi, 1864. *Graphite*

9. Conflict, Conflagration, and Tragedy

FOR ALL THE DRAMA ACCOMPANYING THE SALE of Fort Snelling in 1858, it was a mere sideshow compared to the political theater that had overtaken Minnesota in the previous two years. By the time the army departed the old fort, Minnesota Territory had become a state, and the transition—caught up in national politics—was not easy.

The 1856 election, which brought James Buchanan into office, also saw the first national and local impact of a new political force, the Republican Party. It was made up of an unusual conflation of former Whigs, members of the anti-immigrant American Party (known for their secrecy as the Know Nothings), and members of the Free Soil Party. The Republicans' free-soil and free-labor platforms, which favored homesteaders and opposed the expansion of slavery, found popular support in the "Northwest," including Minnesota Territory. Most of the figures who had dominated the political scene in the territorial period were Democrats. Henry Sibley and Henry Rice, both Democrats, had represented the territory in Congress and had close ties to important party figures, particularly Senator Stephen Douglas of Illinois, who had championed Minnesota's rise to territorial status and to statehood.

Prior to admission as a state, Minnesota Territory had a special election in 1857 to select delegates for a constitutional convention to create a state constitution. Following a bitter campaign tainted with race-baiting and immigrant bashing, neither party had a clear advantage in delegates. After much posturing, acrimony, and stubborn defiance by both parties, including the production of separate versions of the constitution, a final official version was eventually adopted and forwarded to the US Congress for approval.

The Senate took up the matter of Minnesota statehood in early January 1858, but here national politics intervened. The admission of Kansas and its status as a free or slave state remained a hotly disputed cause. Southern senators demanded that Minnesota's admission as a free state could only come after Kansas was admitted to the Union as a slave state. Largely through the efforts of Stephen Douglas, the Senate eventually agreed to consider Minnesota's admission on its own, yet it took another five months for Congress to approve Minnesota's constitution and allow it to become the thirty-second state on May 11, 1858. Henry Sibley, who had barely defeated Alexander Ramsey in the 1857 contest, was sworn in as the state's first governor on June 2. For his part, Franklin Steele had experienced further disappointment. He had entertained high hopes of being one of Minnesota's two US senators along with his old ally Henry Rice, but the Minnesota legislature elected Rice and a smooth-talking, popular newcomer, General James Shields.[1]

The new state's legislators were eager to improve transportation, which would aid development. For as long as five months of the year, ice on the Mississippi effectively cut off steamboat traffic. Minnesota voters promptly amended their constitution to allow a mammoth sale of bonds for rail development, but in the depressed economy of the period, the effort ended as a fiasco of mismanagement and corruption. By 1860, the only concrete results of these efforts were a few surveyed routes with a bit of grading—and a bitterly sarcastic political cartoon that established the state's mascot, the gopher.

And what of the Dakota? In 1858, yet another treaty—really, the forced surrender of reservation lands north of the Minnesota River—was imposed on Dakota leaders who were brought to Washington, DC, to learn of it. They were promised a dollar an acre for the lands, but in June 1860, Congress finally approved only thirty cents per acre, much of which was claimed by traders to repay debts. These terms evoked vehement resentment. There were additional divisions within the community over ways to survive on a reservation that could not support their traditional way of life. The Indian agent enticed individual Dakota families to become "farmer Indians," providing equipment and extra supplies to those who began farmsteads on family allotments—and withholding goods from those who did not. In the rush of the land boom and the chaos of its crash, the newcomers paid little attention to those whose land they had taken.[2]

Through all this, the old fort mostly slumbered, though its new owner continued to have issues with the government. Franklin Steele failed to pay the second installment on the $60,000 he still owed, and on November 11, 1858, the secretary of war directed his department to bring suit against Steele. The suit and subsequent legal actions dragged on for several years, but Steele never paid more than a fraction of the agreed price.[3]

The one public role played by Fort Snelling in these years came in 1860, when the State Agricultural Society chose it as the site for the third annual Minnesota State Fair. Steele, through the Fort Snelling Company, offered the use of the site, quite possibly to promote it as a sort of meeting or event location. The *Stillwater Messenger*, reporting on the society's initial tour of the fort, noted, "Never was the old sentiment of 'turning swords into plough shares, and bayonets into pruning hooks,' more visibly brought to mind. Those old barracks and storerooms, built in the most solid manner and at vast expense, as a place of protection for soldiers and war stores, and abandoned as useless before the irresistible march of settlement, and civilization, and improvement, are now turned into the best of all possible uses—affording place and means for agricultural communion and interchange of progressive ideas of improvement."[4]

The fair itself, held at the end of September, was regarded as something of a mixed success. Commercial agriculture was only beginning; building farms from homesteads took years of hard labor. As one newspaper noted, "The exhibition we should judge to have been more successful in point of attendance, than in display of articles." While most writers admired the buildings, several mentioned that the fort's main problem was one that had plagued it for years. "The greatest difficulty we saw was their inaccessibility," reported the *Stillwater Messenger*, "the boats running from St.

The second state fair, in 1860, was also held at Fort Snelling. Cassius M. Clay, a southern planter and abolitionist, delivered the opening address. *MNHS.*

The ferry across the Mississippi River at Fort Snelling, about 1860. *Library of Congress.*

Paul, and the ferry at the Fort, being inadequate to accommodate the vast number of people on the east side of the Mississippi. On Thursday, ten to twelve thousand persons were in attendance, while the crowd on Friday was quite as large."[5]

While Minnesotans were enjoying the state fair, it was also another election year, one that turned out to be among the most important in American history. The Republicans nominated Abraham Lincoln, a somewhat obscure Illinois politician who had served a short time in Congress and run unsuccessfully for the US Senate against Stephen Douglas. Lincoln was known from his published debates with Douglas, however, and his defeated opponent for the Republican nomination, William Seward, proved a remarkably effective campaigner on Lincoln's behalf. Seward visited the western states of Wisconsin, Minnesota, and Iowa, emphasizing the Republican Party's free-soil policies favoring a generous homestead law and limiting the spread of slavery. A disastrously divided Democratic Party had little answer. Stephen Douglas, one of the two Democratic nominees, should have been popular in Minnesota. He had championed its formation as a territory and promotion to statehood. But Douglas's much-touted "popular sovereignty" policy regarding slavery in new territories had proven a disaster in Kansas, and his party's policies, aimed at appeasing its southern wing by opposing homestead laws and internal improvements, alienated many western voters. In the end, the election was an unqualified victory for Minnesota Republicans, with Lincoln winning the state by a nearly two-to-one majority. Alexander Ramsey was easily elected governor with a strong majority in the legislature. The lone Democratic survivor was Senator Henry Rice, who had not been up for reelection. Democratic governor Henry Sibley had wisely decided not to seek a second term.

For the United States, the election seemed catastrophic. In the four months between the vote and the new president's first day in office, momentum for secession grew among southern states, and the Buchanan administration seemed to have little ability or even inclination to prevent it. In fact, some believed that federal officials like Secretary of War John B. Floyd were openly sympathetic to the southern cause.[6]

Minnesotans could follow the crisis as it evolved. Telegraph connections reached St. Paul and Minneapolis late in 1860, and the proliferation of newspapers in Minnesota in the previous decade meant that parties could deliver the news with their own perspectives. Though such conflicts over slavery were alarming, they had come before, in 1820 and 1850, and many probably believed that some compromise would again settle the matter. On April 13, 1861, that notion was laid to rest when news arrived that South Carolina troops had fired on Fort Sumter, the federal post guarding Charleston harbor. "The ball is opened, War is inaugurated," wrote the *St. Paul Pioneer and Democrat*, though it still expressed some faint hope for compromise. The *Winona Daily Republican* had no such illusions: "That collision which has for months been impending is at length forced upon the government by the Southern traitors. War has actually broken out. The die is now cast, and there must be no turning back until the fact is ascertained, at the cannon's mouth, that we still have a Government."[7]

President Lincoln soon called for volunteers to put down the rebellion. At the time fighting broke out at Fort Sumter, Governor Ramsey happened to be in Washington lobbying for patronage positions for Minnesota Republicans. He took the opportunity to gain Minnesota the distinction of being the first state to commit to raising a regiment for national service. Secretary of War Simon Cameron forwarded Ramsey's proposal to President Lincoln, who quickly accepted it. Under the military system of the time, the president could call for volunteers from the various states to augment the strength of the regular army. Though state militia companies frequently formed parts of these volunteer regiments, the units themselves were largely made up of new volunteers. Each state, through its adjutant general, commissioned citizens to raise companies of recruits, who would then report to a gathering spot known as a volunteer rendezvous. Here these volunteer companies were organized into regiments and received basic military instruction and equipment before being taken into federal service. In Minnesota's case, there was one very obvious location for gathering volunteers: Fort Snelling. The old post offered a well-known place to house and drill new recruits positioned near water and road transportation.

Recently appointed state Adjutant General John B. Sanborn met with Franklin Steele and obtained his consent to allow the State of Minnesota to use the fort without charge. Before the end of April, new recruits for the 1st Minnesota Volunteer Infantry were arriving daily, but the post was ill equipped to receive them. In the two years before the sale of the fort, the army had done little in the way of maintenance or repair, and Franklin Steele had not invested much in keeping up the old buildings, some of which had endured more than forty years of exposure to Minnesota's climate. The business generated by the renovations was welcome in the tight local economy.[8]

At first, conditions at the fort were somewhat chaotic. The new recruits of the 1st Minnesota were exuberant and anxious to see action, but discipline was slack. Civilian visitors came and went, creating a carnival air. On the day late in April when the regiment was mustered into federal service, the post quartermaster, First Lieutenant Thomas Saunders of the 3rd US Artillery, encountered an unusual situation, as he later related in a letter to Quartermaster General Montgomery Meigs (see sidebar).

Lieutenant Saunders had served previously at Fort Snelling as an officer in Battery E of the 3rd Artillery. He remained after the unit was transferred to Fort Leavenworth in 1857, acting as the army quartermaster in St. Paul in charge of shipping supplies to the frontier posts at Forts Ripley, Ridgely, and Abercrombie (built in 1858 on the west bank of the Red River). A Virginian by birth, he refused a commission in the Confederate Army in 1861 and lost claim to his family estate as a result. Following the sale of Fort Snelling, Saunders worked out of the warehouse the army rented near the upper levee in St. Paul. He shifted operations back to the fort when it came into use as a volunteer rendezvous. He was among the few regular army officers immediately at hand, and he worked closely with state officials.[9]

Fort Snelling,

Minnesota
September 14, 1861
Brig. Gen. M. C. Meigs
Qrt. Master Gen. U. S. A.
Washington, D. C.

Sir:

I have the honor to enclose herewith, an estimate of the cost of some repairs, absolutely needed at Fort Snelling, Minn, and will give as briefly as possible, my reasons for making the estimate.

When the 1st Regiment of Volunteers from this state, was ordered to assemble at the Fort for the purpose of being mustered into service, they took with them a six pounder, and fired a salute within the walls, a proceeding never allowed before.

The concussion broke a good many sash and glass, and as the weather is getting quite cool, it is necessary, in order to make the quarters habitable, to have the sash and lights put in again.

I was present when the damage was done, and tried to dissuade the men from firing within the fort but was told it was none of my business, they would fire as much as they pleased.

The men had not been mustered in at the time, and could not be controlled.

I am sir,
Most Respectfully,
T.M. Saunders
1st Lieut. 3rd Art. A.A.Q.M.[i]

In May, to the annoyance of some of the new recruits, companies of the 1st Minnesota were sent out to Fort Ridgely, Fort Ripley, and Fort Abercrombie to relieve regular troops for duty elsewhere. This situation proved only temporary, as the entire regiment was ordered to Washington, DC, on June 12. By the twenty-second of the

Officers of the 1st Minnesota Volunteer Regiment in front
of the commandant's quarters, May 1861. *MNHS.*

month, the dispersed companies had reassembled at Fort Snelling, and the whole regiment boarded steamboats for the journey east. Meanwhile, companies from the 2nd Minnesota, then forming at the fort, had taken the place of the 1st Regiment's men at Forts Ridgely, Ripley, and Abercrombie. Thus a pattern was established that would exist for the next year. Regiments mustered in, then were sent to relieve older regiments on frontier duty. The older regiments then departed to join other commands, while more new units were being organized at Fort Snelling. In all, Minnesota furnished eleven regiments and one battalion of infantry, two companies of sharpshooters, two

regiments and two battalions of cavalry, three batteries of light artillery, and one regiment of heavy artillery to the Union cause. In addition, several thousand men who were drafted into the service principally served as replacements in established Minnesota units. In all, approximately 24,000 men and at least two women served from the state. Record-keeping was not exact in the mid-nineteenth century, and a significant number of men enlisted in one unit and later reenlisted in another, so the precise number is unknown, but the total represented a significant percentage of the male population eligible for service.[10]

Nearly all of these soldiers came through Fort

Snelling during the course of their military service. Companies were usually formed within the state's new towns. A reasonably well-known and well-connected community leader, often someone who had held an elected office, recruited local men to form a company. When enlistment neared or reached sixty men, the minimum number for a company, they traveled to Fort Snelling as a unit. There, recruits were enrolled again and began learning the basics of military instruction and discipline. When ten companies had gathered—the number required—the regiment was organized. Company officers were elected, and the regiment's "field officers"—the colonel, lieutenant colonel, major, and quartermaster—were appointed by the governor. Then the new regiment was mustered in, and the men became soldiers in the US Army.

Unlike previous troops at Fort Snelling, these were truly citizen soldiers: farmers and artisans, clerks and laborers. Most could read and write to at least some degree and had an understanding of the issues surrounding the war. They volunteered for a variety of reasons: to preserve the union, to oppose slavery, or just for the experience—"to see the elephant," to quote a commonly used expression of that time. On his way to Fort Snelling in October 1861, Thomas Christie explained his reasons for enlisting in a letter to his father that echoes what many young men of his generation were probably feeling:

> Now you know, father, that you would enlist if you were in my place. You have taught me to hate Slavery and to love my Country. I am only carrying out these principles in coming now to the help of the Country when she is attacked by a Slaveholders' rebellion.
>
> I shall not deny that motives other than strictly patriotic have had an influence upon me; but I don't think these other motives are wrong. I do want to "see the world," to get out of the narrow circle in which I have always lived, "to make a man of myself," and to have it to say in days to come I, too had a part in this great struggle. . . . I feel

sure, even as I write, that you will not only give me your blessing—but that you will even be glad to have your son enrolled among the defenders of the Union. But whether that be so or not, I must go.[11]

By the summer of 1861, the numbers of recruits had outstripped the capacity of the stone barracks to hold them. A tent encampment was created on the prairie just northeast of the fort, and new buildings and enclosures would soon follow.

Meanwhile, the old barracks were renovated and improved. Initially, many of these improvements were done at state expense, but after 1861, the federal government assumed the costs. The economic impact on the local economy of this spending for materials, supplies, and labor must have been considerable.[12]

Just supplying food for the 1st Minnesota was an eye-opening challenge, as the *Chatfield Democrat* reported, "GOOD FEEDERS.—The amount of food cooked each day for the 1st Regiment of Minnesota Volunteers now at Ft. Snelling, has been 40 bushels Potatoes: thirteen to fifteen hundred pounds of beef: 210 pounds pork 1500 loaves bread 25 gallons coffee: 2-10 pounds sugar: 110 pounds butter: one barrel dried apples, &c."[13]

During the course of the next four years, hundreds of contracts would be let by the army in support of the troops at Fort Snelling. The military purchased not only food, but also the ever-necessary wood, both for fuel and building materials, and the transportation needed to move both supplies and people. The requirements of Fort Snelling and the general war effort stimulated growth of what became the region's economic engines for the rest of the century: milling of lumber and flour, agricultural production, and transportation. Companies like Washburn's Minneapolis Mill Company, Davidson's La Crosse and Minnesota Steam Packet Company, Burbank's North-Western Express Company, and the St. Paul and Pacific Railroad (which actually built and operated ten miles of track between St. Paul

Company E of the 8th Volunteer Infantry at Fort Snelling, 1862. *MNHS.*

and St. Anthony) all received important direct and indirect stimulus from military and government contracts.[14]

Through the first half of 1862, the fort functioned primarily as a volunteer rendezvous, housing, feeding, and equipping new troops on a scale never seen in its former life as a frontier garrison. The February 25, 1862, *Pioneer and Democrat* reported:

> The reports at Fort Snelling on Sunday morning showed about 1,200 troops at the post, composed of five companies of the Fourth Regiment, the Second Battery of Artillery, and companies of the Fifth Regiment—the latter amounting already to 600 men. By an order read on dress parade on Sunday evening, Col. Sanborn assumed military jurisdiction over all that portion of the old military reserve within one mile of the walls of the fort. This cause was necessary for obvious reasons, in order to preserve a proper degree of order and discipline among the troops. We have never seen Fort Snelling since its first occupation last April, in a better state of military order than it now is.

There were also suggestions that Fort Snelling might serve as a prison camp for the increasing number of Confederate captives from Ulysses S. Grant's campaigns on the Mississippi, or a regional hospital for convalescing soldiers from the western theater of the war. But events during the summer and autumn of 1862 would drastically shift priorities.

While the war in the South raged on with shocking casualties and no end in sight, most white Minnesotans had given little attention to the Dakota or the conditions on their reservation. The winter of 1861–62 had been harsh, leaving many of those living on the reservation in a state of starvation. To make matters worse, the annuity payments were late. The federal government was focused intently on the war with the Confederacy and had little gold currency to spare for the promised payments. Many of the traders refused to extend credit, while Indian Agent Thomas Galbraith, a recent Republican appointee, proved unable or unwilling to resolve the problems. Though some in the Indian agency and military at Fort Ridgely had expressed concern over the increasing desperation among the Dakota and growing tensions with the traders, little was done to defuse the situation.

On August 16, 1862, a small group of men returning from an unsuccessful hunt came into conflict with a farmer near the town of Acton. They killed him, his wife and daughter, and two others. The young men sought refuge at Shakopee's village, and the more traditionally minded leaders there determined that rather than turn over the men responsible for the killings, as they might have done in the past, they would launch an all-out war on the whites. These were mostly young men, members of the traditionalist Soldiers Lodge, driven by the desperation of their people's hunger as well as a desire to regain their lost lands and the way of life that had gone with them. They also believed that the Civil War had left the whites vulnerable to attack. Knowing that many Dakota, particularly those who had taken up farming, would not support such a war, they hoped to persuade the most prominent of the Dakota leaders, Little Crow (Ta Oyate Duta), to lead them. Little Crow's son, Place of Refuge (Wowinape), stood beside him and later provided an account of this meeting. Little Crow knew that any war would be hopeless and that he was dealing with young men who had no firsthand knowledge of the United States beyond the colonists living near the reservation. But the younger men had already convinced themselves that war with the whites was possible, even desirable, and rejected his advice. They accused Little Crow, who had long been a prominent spokesman for the Mdewakanton, of cowardice. Enraged, the leader accused them of acting like children. He pointed out that the whites were as numerous as "the locust when they fly so thick that the whole sky is a snow storm." Though the Dakota would die "like the rabbits when the hungry wolves hunt them in the Hard Moon," he would lead them.[15]

Within a few days, they had attacked the Upper Sioux Agency, killing twenty of the traders and clerks who had antagonized the Dakota. As the fighting spread, Dakota warriors began to attack white families. Many were killed as they tried to flee to the safety of towns, while some of those towns, like New Ulm, Hutchinson, and Forest City, were attacked by war parties. But even in the midst of all this bloodshed, people who had treated the Dakota with respect or had kinship ties were often spared and made prisoners. The Dakota community, so tightly bound by relationships, was split. Sisseton and Wahpeton leaders argued against the war, although some of their young men joined; Mdewakanton men, whose lands had first been invaded and who had been the furthest displaced in the removals, were more inclined to fight. Many white captives found protection in the lodges of those who were not fighting. Other Dakotas warned settlers and opposed the fighting.[16]

In the next six weeks, more than six hundred people, sometimes whole families with small children, were shot down or bludgeoned to death. The white community reacted with a mix of panic and fury. The already lurid reality was enhanced by exaggerated stories of atrocities spread by terrified refugees. The accounts were repeated in the local press and picked up by national publications. Feeling a sense of outrage and betrayal that a people deemed destitute, dependent, and largely harmless had attacked whites on such a large scale, the white community demanded first protection—but above all, retribution.[17]

As news of the killings reached St. Paul, Governor Ramsey appealed to the federal government for assistance, but realizing any help from that quarter would take time to arrive, he knew immediate local action was needed. On August 19, the governor approached his former political antagonist, Henry Sibley, asking him to command an expedition of the state's militia against the Dakota. The former-fur-trader-turned-politician knew little of military matters but had decades of experience interacting with the Dakota, as well as kinship ties. Sibley agreed, providing he had a free hand to act without interference from the US military or politicians. Ramsey acquiesced to this request. On the next day, he gave Sibley a commission as a colonel in the state militia and command of the only state troops available, four newly recruited companies of the 6th Minnesota Regiment, who were at Fort Snelling waiting to be mustered into federal service. The "Indian Expedition," a troop of disorganized, undisciplined mounted volunteers from St. Paul and Minneapolis carrying whatever supplies could be gathered together, set off up the Minnesota River for Fort Ridgely. Sibley wrote privately to his wife, "a greener set of men were never got together."[18]

International Hotel, 115 East Sixth Street, St. Paul, 1865.
Photo by Martin's Gallery, MNHS.

Authorities at Fort Snelling began sending supplies and reinforcements to the expedition as quickly as possible. The most important of these were 250 men from the 3rd Minnesota. Some months before, this well-trained but unlucky regiment had been controversially surrendered by their commanders to Confederate forces at the Battle of Murfreesboro, Tennessee. Unable to take them all as prisoners, the Confederates had "paroled" the enlisted men, requiring them to sign an oath stating they would not fight against Confederate forces until exchanged for a like number of Rebel prisoners. The men of the 3rd had expected to wait out their parole at Fort Snelling or guarding frontier posts, but the outbreak of this war with the Dakota gave them a more active mission. From Sibley's perspective, he was fortunate to have these trained, disciplined soldiers, and they formed the core of his command.

Another important source of aid came to Minnesota from Washington. On September 6, the War Department, in response to Governor Ramsey's repeated pleas, created the Military Department of the Northwest, appointing Major General John Pope as commander. Pope had led the Union Army in Virginia into a crippling defeat at the Second Battle of Bull Run. Unpopular with the Army of the Potomac, he was politically disposable, and the Northwest was a conveniently distant place for him. Though a blustering self-promoter, Pope was nonetheless an effective, experienced administrator and knew the region well. As a young officer in the Topographic Corps, he had taken part in a survey of the region in 1849 and had thought well enough of the area to purchase lots in St. Anthony, which he later sold to Franklin Steele.[19]

General Pope and his staff arrived in St. Paul on September 15, establishing their department headquarters at the International Hotel. This new military jurisdiction, including the states of Minnesota, Wisconsin, and Iowa, along with the Dakota and Nebraska Territories, gave General

Pope wide authority to command troops and resources throughout the region. Even before arriving in Minnesota, Pope had asked the governor of Wisconsin to send four newly formed regiments from that state to Minnesota, and he had requisitioned 2,500 horses to mount as many of Sibley's troops as possible. This new command structure also marked the beginning of a new direction in the history of Fort Snelling. Though it continued as a rendezvous for volunteer units—and, later, draftees—the post increasingly became the supply and transportation hub for military operations reaching from the Mississippi River to the Yellowstone.[20]

Upon arrival in Minnesota, one of Pope's first acts was to write Sibley a long letter confirming his confidence in the former governor and assuring him of support, a gesture that Sibley valued as he was being mercilessly attacked in the local papers for his cautious and slow-moving campaign. Soon enough, Sibley's approach was vindicated. By the end of September, his expedition had won a battle at Wood Lake and soon accepted the surrender of a camp of 150 lodges of Dakota and people of mixed ancestry near the Upper Sioux Agency. These included a significant number who had not taken part in the fighting and had opposed going to war at all. This "Peace Party" arranged the release of the hundred or so whites they held "captive" to protect them from harm. Its leaders clearly understood, through intermediaries, that Sibley promised to punish only those who had killed white settlers; soldiers would be treated as prisoners of war. Little Crow, along with several hundred of his followers and their families, escaped to the north and west.[21]

Sibley named the place Camp Release and hoped, he told his wife, to return to civilian life himself. While still at work arranging matters at Camp Release, however, he was honored by President Lincoln with a promotion to the rank of brigadier general of United States Volunteers.

This elevation seems to have changed Sibley's outlook as he faced the complex problem of what to do with the Dakota captives. Resisting public outcries for immediate retribution, the newly promoted general took a more legalistic approach. Any of the men believed to be combatants—virtually all the men of fighting age—were separated from the others. An ad hoc military tribunal heard their cases in trials, some lasting only a few minutes. In all, 303 men who admitted being present at battles were condemned to hang, sentences that Sibley was all too ready to carry out and General Pope eager to confirm. The latter recognized, however, that the president had the ultimate authority in such matters, so the final decision on the trials and convictions was left to President Lincoln. Pope also instructed Sibley to send the noncombatants—about 1,700 women, children, and elderly—to Fort Snelling for the winter. In fact, Pope initially ordered that all Dakota be moved to Fort Snelling, where the executions would take place. Fort Snelling was the most expedient place from which to expel the Dakota from their homeland.[22]

Sibley had all of the imprisoned men moved to the Lower Sioux Agency, where supply and communications were easier, and then the men were shackled and hauled to Mankato in wagons. The women, children, and elderly packed up their camp and made the long trek by road to Fort Snelling. Both groups suffered at the hands of vengeful settlers en route, despite the army's attempts to protect them. Though many of the veteran soldiers were disciplined and obeyed orders, some of the newer volunteers and militia were not as reliable. Guarding the male prisoners proved particularly difficult, as much of the anger of local residents and the citizen militias was focused on them. When the prisoners were first held in a camp near Mankato in early December 1862, a mob of citizens from Mankato and other nearby towns attempted to attack the camp with the intent of killing all the Dakotas and anyone who tried to stop them. Only the firm command of the prison guard by veteran colonel Stephen Miller prevented what might have been a horrific mass murder. The journey of the Dakota families also proved difficult. A mob attacked their caravan in Henderson, beating and injuring many of them, including an infant who died soon after. A Dakota account of their march tells of the killing of an elderly woman by soldiers when a cart became stuck on a bridge "near New Ulm or Morton."[23]

On November 13, 1862, the women and children arrived at Fort Snelling, where they encamped on the bluff, near the former Indian agency; a few days later, they were moved into a new enclosure built at the base of the bluff below the fort, near the steamboat landing. This was a sad and bitter return to Bdote, the sacred place at the center of the Mdewakantan homeland, the site of many ceremonies and gatherings. They had come with their wagons, horses, cattle, and tipis. The wagons and animals were soon stolen or sold for what little they could get to help support themselves. The tipis were their shelters for the winter. Conditions were crowded, with nearly 1,700 people living on two to four acres, with little sanitation. Young women were harassed by off-duty soldiers, despite the army's attempts to stop the practice; Dakota oral history tells of rapes committed in the camp, and local newspapers reported that a woman gathering firewood outside the camp was brutally raped. The psychological strain for the captives, separated from husbands, fathers, and sons, dislocated and uncertain of their future, must have been terrible.[24]

One Wisconsin soldier, Chauncey Cooke, newly arrived at Fort Snelling, left a vivid account of what he saw in the camp in late November 1862.

The concentration camp below Fort Snelling, 1862–63.
Photo by Benjamin Upton, MNHS.

They are a broken-hearted, ragged, dejected looking lot. . . . Papooses are running about in the snow barefoot and the old Indians wear thin buckskin moccasins and no stockings. . . . I lifted up the flaps of a number of their tepees and looked in. Every time I looked in I met the gaze of angry eyes. Nearly all of them were alike. Mothers with babies at their breasts, grandmothers and grandsires sat about smoldering fires in the center of the tepee. . . . The white man's face was their hate and their horror and they showed it by hate in their eyes and their black lowering brows. Why shouldn't they? What have they done? What was their crime? The white man had driven them from one reservation to another. They were weary and broken hearted and desperate at the broken promises of the government. And when they took up arms in desperation for their homes and the graves of their sires, they are called savages and red devils. When we white people do the same things, we are written down in history as heroes and patriots. Why this difference? I can't see into it.[25]

Few Minnesota residents were as sympathetic as Private Cooke. White survivors, too, were forever separated from loved ones who had been killed. But the Dakota lost everything—and the whites would, as they had long anticipated, end up with the land. The prevailing opinion, reflected in the press, called for removal of all Indigenous people (including the Ojibwe and Ho-Chunk), whether involved in the fighting or not, and the punishment, preferably execution, of those who had attacked whites.

Following the surrender of the Dakota at Camp Release, in October General Pope had declared fighting over for the remainder of the year, though he personally desired an offensive against the Dakota for political reasons: he was under pressure from Minnesota Senator Henry Rice (who reportedly wanted Pope's job), and he very much desired to restore his military reputation. When rumors of a possible peace treaty with the Dakota not involved in the conflict reached Jane Grey

Swisshelm, the editor of the *St. Cloud Democrat*, she wrote, "'A permanent peace with the conquered Sioux,' is it? Not by the Lord of Heaven is the vow of 200.000 American citizens on the soil of Minnesota, and not on any soil adjoining it, except, through the death of the guilty and the removal of every surviving shred and remnant of this hated race. This is no longer the petition of the people of Minnesota. It is their demand—a demand which they will exact to the last retributive drop of Indian blood."[26]

Such sentiments were common in the months immediately following this war-within-a–Civil War. "Our course then is plain," Governor Ramsey said to a special session of the Minnesota legislature on September 9. "The Sioux Indians of Minnesota must be exterminated or driven forever beyond the borders of Minnesota."[27]

As often happens in war, disease follows the displacement of refugees, and this was the case at Fort Snelling as well as the nearby towns. Measles, mumps, and influenza, along with associated pneumonia, ravaged captives, soldiers, and refugees alike. For the Dakota, weakened by the months-long ordeal, the diseases proved especially hard and particularly fatal to the very young and the elderly. The exact death toll is unknown, perhaps because families buried their dead within the camp to protect their bodies from desecration. In a letter to the *St. Paul Daily Press* dated December 11, 1862, missionary Stephen R. Riggs recounted, "Many things have been attempted, and some of them done, which make me feel greatly ashamed of our people. Since the Dakota camp has been placed at Fort Snelling, quite a number have died of measles and other diseases. I learn that their buried dead have, in several instances, been taken up and mutilated." Official records count 165 deaths in this period, one-tenth of the population of the camp. The measles outbreak was also virulent among the troops at Fort Snelling, sending many to the post hospital and

Was the Fort Snelling camp a "concentration camp"? Its inmates were in judicial limbo, not accused or convicted of any crimes, but not exactly free to go. Given the intense hostility of the local population, leaving could be dangerous. The Dakota were given rations similar to but slightly smaller than army rations. Soldiers regularly came into the camp to patrol it for unwanted civilians or other soldiers, while some like Franklin Steele took financial advantage of the situation.[ii]

Though regularly referred to at the time as an "internment" camp, the Fort Snelling site accurately fits the description of a "concentration camp" as defined by the US Holocaust Memorial Museum: "a camp in which people are detained or confined, usually under harsh conditions and without regard to legal norms of arrest and imprisonment that are acceptable in a constitutional democracy." The term was not used in the 1860s, but it originated three decades later in two instances where imperial powers sought to control insurgencies by rounding up civilian populations. During the Cuban revolution, Spanish military officials "reconcentrated" a third of the Cuban population in overcrowded camps, rife with disease and malnutrition, by 1898; some 150,000 died in these places. Subsequent coverage of this atrocity in American newspapers contributed to the US war with Spain. In South Africa during the Second Boer War in 1900, the British rounded up nearly 200,000 Boer civilians into enclosed camps where thousands died from disease. The technique of confining civilian populations regarded as troublesome continued to be used even by US forces in the Philippines. The use of the same term, *concentration camp*, was most notoriously applied in Nazi Germany before and during World War II, both for camps containing political prisoners and for the death camps of the Holocaust that were deliberately designed to murder people on an industrial scale. There the concept of confining "undesirables" reached a particularly horrific culmination, leaving the term *concentration camp* with a meaning even darker than its original form. It is not a label to be applied lightly.[iii]

Using a twentieth-century definition to describe a mid-nineteenth-century phenomenon might seem unmerited, but we often use terms invented after 1900 to discuss earlier events: "superpower," "gender role," or "ethnic group," for example. The description in its original sense is indeed accurate. Though the Fort Snelling camp was decidedly not an "extermination camp" in the World War II sense, it had the same effect as the Cuban and Boer War concentration camps in terms of human misery. In both instances, defenseless civilians, held without being convicted of any crime in close confinement in unsanitary conditions, died in large numbers from disease and malnutrition. In the case of the Dakota, one might ask, *What were the alternatives?* Could the army leave these women, children, and elders on the prairie to starve or endure the vengeance of the infuriated settlers? Certainly these were not good alternatives, but even by nineteenth-century standards, the camp's size and location—a cramped enclosure on wet bottomland with poor sanitation—indicate a kind of malign neglect. One can also wonder what might have happened if the Dakota at Camp Release had been given rations and sent on their way to find their relatives among the Yankton. A politically impossible option, perhaps, but maybe a more humane one.[iv] ⟨⟩

resulting in a number of deaths. St. Paul, now crowded with refugee settlers, was also hard hit.[28]

—

The uncertainty and fear among the prisoners led to something unprecedented among the Dakota, both at Mankato and in the Fort Snelling concentration camp. For decades, white missionaries had tried to convert Dakota people to Christianity, with very little success. For white officials, such conversion was deemed essential for the Dakota or any other Native people to become "civilized" and therefore acceptable to European American society. Now, in desperation, many in the camp turned to the missionaries for help and accepted baptism and confirmation. A number of missionaries, including Episcopal Bishop Henry Whipple, one of the few prominent whites to openly sympathize with the Dakota, preached and proselytized in the camp. Working with

Whipple was the Reverend Samuel Hinman, who lived within the camp. During his stay, Hinman was attacked one night by "white roughs from St. Paul" who beat him severely. Nearby clergy also ministered at the camp, including Father Augustin Ravoux from the Roman Catholic parish at Mendota and the veteran missionaries Gideon and Samuel Pond from Bloomington. A number of the converts learned to read and write, partly out of their desire to communicate with loved ones held at Mankato.[29]

⸺

While the missionaries were saving souls, others, as usual, had their keen eyes set on profits. Franklin Steele had already benefited from supply contracts feeding the army and the Dakota prisoners at Fort Snelling. He was also allowed to run a sutler's store within the camp. Steele and a few others, including Henry Rice, were aware of another potential "treasure" there, the so-called half-breed scrip. This was a convoluted story. In the 1830 Treaty of Prairie du Chien negotiations, Chief Wabasha asked that a tract of land be set aside for the Dakota's relatives of mixed ancestry. An area west of Lake Pepin of around 320,000 acres was identified, and Article 10 of the treaty gave the president of the United States authority to allot tracts of 640 acres to each "half breed." Many of those in the fur trade, including Alexis Bailly (to whom Wabasha was heavily in debt), were themselves of mixed descent and had Dakota wives and many children. They might theoretically be "entitled" to thousands of acres per family.[30]

For some years, nothing happened with this land, and it remained unallotted, despite lobbying efforts from Steele, Sibley, and Stambaugh. The traders even tried to substitute a payment of $150,000 into the 1851 Treaty of Mendota in place of these land claims. In 1853, Henry Rice, by then a US delegate for Minnesota Territory, introduced a congressional bill to allow scrip to be issued for up to 650 acres of "unclaimed and unsurveyed" federal lands in return for the "mixed-blood" claims. These scrip were not supposed to be transferable, but through clever use of powers of attorney, the land might be transferred and sold, with the final location of the land and name of the deed holder left blank, thus making the scrip potentially quite valuable.[31]

More than half the scrip was sold prior to 1857, much of it to Steele. The panic of 1857 and then the Dakota war reduced the value of the Minnesota lands he acquired in this way, but Steele was well aware that many people of mixed ancestry had been swept up in Sibley's 1862 campaign and found themselves crammed into the Fort Snelling concentration camp. In particular, the descendants of Joseph Renville and members of missionary Stephen Riggs's Hazelwood community were anxious to find a way out of the camp. Some led by Gabriel Renville agreed to act as scouts for Sibley and later sold scrip to Steele; others sold some eight thousand acres' worth of scrip to Franklin Steele for cash before they were sent off to Crow Creek.[32]

By the fall of 1863, Steele, working with several partners, had amassed scrip for some 15,000 acres of federal lands. But rather than selling it in Minnesota, where depressed land prices reduced their value, Steele, with his partner William Chapman acting as agent, sold the scrip in Carson City and Virginia City, Nevada, where the fabulous Comstock Lode had sparked a major mining boom with accompanying land speculation. Chapman sold the scrip for $10 to $15 per acre. Back in Minnesota, Steele and his other partners helped provide the necessary paperwork in the form of powers of attorney and land patents to complete the sales. The cash bonanza from these sales financed still more speculation and investment, much of it in Minnesota, including the Northern Pacific Railway and the Northwestern National Bank of Minneapolis.[33]

From the *St. Paul Daily Press*, April 24, 1863: "As the steamer approached Fort Snelling through the tortuous channel of the Minnesota, the red blankets of the Indians on board were discovered, while still miles away, by some of the Sioux under Government protection at Fort Snelling, who were fishing on the bank of the river. The news that something interesting was coming was telegraphed from one strolling Indian to another by shouts and waving blankets, and in a few minutes reached the Indian encampment at the Fort. The whole crowd of Indians, men, women and children poured out of their encampment, watching the progress of the steamer with strained eyes and eager expectancy. As she approached nearer they too caught sight of the blanketed forms and the easily recognized Sioux head dress. They readily divined that these were the friends, or such as survived the executioner, from whom they had been separated last fall at Fort Ridgley. Moved by a common impulse the whole throng of eager and anxious spectators rushed to the levee to await the arrival of the steamboat and perhaps to greet and rejoice the husbands, brothers and sons whom they had scarcely dared to hope they should be allowed to see alive again. . . . The forty-eight unconvicted Indians . . . were quickly put ashore, the plank drawn in, and the boat pushed off again on its voyage, when the poor Indian women, the wives and mothers, whose husbands and sons were not in the released number, saw the boat moving off without discharging their relatives, with frantic dismay. . . . The whole vast crowd of savage forms writhed in the agony of disappointment, and a wail of grief went from hundreds of shrill, wild voices which it was heart-rending to hear. The poor creatures hurled themselves on the ground, and pulled their hair, and beat their breast with the anguish of the sudden revulsion from hope to despair. . . . Capt. Crooks called the poor despairing creatures together in council, and through an interpreter, explained to them that the Indians on board the boat were not going to be hung, but to be carried down river to a more convenient place, and placed under another guard. This assurance calmed them somewhat, but air was still filled with their lamentations as if they mourned the dead."

In March 1863, rising water and unsanitary conditions in the camp prompted the fort's commandant, Lieutenant Colonel William Crooks, to order the encampment moved to a location on high ground outside the fort. In April, the forty-eight Dakota men who were acquitted by the government were taken to Fort Snelling to join the others interned there. On the same boat from Mankato were some 265 others sentenced to prison terms at Davenport, Iowa. To avoid demonstrations or attacks by whites, the army had arranged this transfer secretly. The result for the Dakota families at Fort Snelling was a heartbreaking repetition of separation and anxiety.[34]

By March 3, 1863, the federal government had decided on the removal of the Dakota who were held at Fort Snelling. Congress passed the Dakota Removal Act, which mandated their relocation to an area near the Missouri River known as Crow Creek, now in South Dakota—and also provided for the sale of the reservation lands to settler-colonists. Crow Creek, chosen for its proximity to Fort Randall, was a miserable location for a farming community, with bad water and poor soil. On May 4, 1863, the first group of 547 were packed into the steamer *Davenport* for the long journey. On the way downriver, the boat stopped briefly at the St. Paul levee to take on freight. A crowd gathered there and some, "led by a soldier who was wounded at the Battle of Birch Coolie, commenced throwing boulders at the Indians and as they were so closely packed upon the boiler deck . . . it was impossible for them to escape the missiles." The attack was stopped when the troops escorting the Dakota, Company G of the 10th Minnesota, threatened the crowd with their

Fort Snelling, 1865. *Photo by Joel Whitney, National Gallery of Art; Clinton and Jean Wright Fund.*

bayonets. "While in common with most of the citizens of this State, we do not feel particularly partial to the Indians, we cannot consider this attack upon defenseless women and children other than a gross outrage," wrote the editor of the *St. Paul Press*. For those on the *Davenport*, leaving their ancestral homeland for the last time, it was a bitter final departure. The remaining Dakota prisoners left the stockade on the following day on the steamboat *Northerner*.[35]

These were not the last Dakota to live in the stockade. A number of Dakota men had signed on to act as scouts for General Sibley's 1863 "Punitive Expedition" against the Dakota and Lakota bands beyond the Red River. Their families lived for a time in the camp on the bluff.

But the Dakota were not the only people to be expelled from Minnesota. The Ho-Chunk, who had lived since 1851 near the Dakota on similarly rich and coveted farming lands, had not joined the fighting. Yet their peaceful inaction did not save them. The war served as a pretext for the people of neighboring Mankato to have the Ho-Chunk removed in order to gain access to their reservation lands. Two thousand Ho-Chunk men, women, and children were also brought to Fort Snelling, where they stayed briefly before boarding steamboats for exile at Crow Creek.[36]

———

Further developments took place in the military organization of the region in the later part of 1862 and early 1863. General Pope still remained in command of the Department of the Northwest, but his headquarters moved to Madison, Wisconsin, and later Milwaukee, where the department's work would be facilitated by better railroad connections. Sibley was given command of the District of Minnesota, officially headquartered in St. Paul's International Hotel. Depending on the season, there were regular stage or sleigh routes run by the Burbank Express Company between the town and the fort. Boats often traveled to the fort as well, carrying both supplies and passengers.

Civilians were regular visitors at the post, particularly in the early years of the Civil War, when military reviews were a splendid novelty, and most believed the war would be over in a matter of months. But throughout the war, people went up to Fort Snelling to visit relatives, gawk at Dakota prisoners, or carry out business. There were also more civilians than ever working on the post. In June 1863, the forty-six civilian workers included laborers and mechanics. By September, the quartermaster department listed clerks, a "foreman in the barn," thirteen teamsters, thirty laborers, two blacksmiths, ten carpenters of various pay grades, one mason, and a harness maker, among others. The numbers grew still greater in 1864, when the April returns show over ninety teamsters, the usual clerks and others, and additional bakers and farriers (blacksmiths specializing in making shoes for horses and mules).[37]

All of this activity and large expenditure of government funds expanding the post was in support of a major military expedition against the western bands of Dakota and Lakota as well as for protection of the new routes to the gold fields of Idaho and Montana. Little Crow's remaining forces were relatively weak, no more than a few hundred warriors. He had tried to persuade others of the Oceti Sakowiŋ who had stayed out of the fighting in 1862—the Teton bands and particularly the Yankton—to join him, but to no avail. Yet General Pope believed that thousands of warriors were gathering on the plains, eager to attack white settlements, and that the best course of action was to attack. He expressed his views to the *St. Paul Pioneer and Democrat* in November 1862. He did not merely intend to punish those involved in the August attacks but "to seize and dispose of all the Indians upon whom we can lay our hands." His ultimate objective was to make "the whole region to the Rocky Mountains . . .

opened to immigration, travel and settlement." To accomplish this, he planned for two coordinated offensives into the region. General Sibley was to lead a force of around 3,300 troops, including infantry, cavalry, artillery, and a long supply train supported from Fort Snelling, up the Minnesota River Valley to the Red River, then overland to the region around Devils Lake, and from there southwest to the Missouri River. Here he would rendezvous with a second column of cavalry and artillery commanded by General Alfred Sully that was to make its way from Council Bluffs, Iowa, along the Missouri River. Somewhere along the way, the two columns of troops hoped to trap a large number of Dakota and Lakota.[38]

Some in Minnesota were not enthusiastic about General Pope's plan. During the winter of 1862–63, a line of forts and outposts had been established on Minnesota's western frontier, some built by local militia and others by the military, to offer local protection and patrol the regions between posts. There was concern that Pope's expeditions would remove troops from these outposts and more established places like Fort Ripley and Fort Ridgely. There was also more mundane concern over the potential loss of income for farmers near these posts, who were profiting from selling food and forage to the army. Even among the troops, there were those who thought the whole effort "a humbug," a senseless campaign against an already-defeated foe, while much more important work remained to be done fighting the Rebels in the far-off South.[39]

Sibley's troops set off on June 16, 1863, making an arduous overland march for nearly a month before finally fighting an indecisive battle on July 24 with Dakota and Lakota warriors at a place called Big Mound. Rather than a war party, Sibley had found a large hunting camp with many women and children as well as hunters on their annual buffalo hunt. Most of these people had not been involved in the previous year's conflict.

While their families fled from the soldiers, the warriors made a delaying stand against Sibley's men. Over several days, the US troops continued pursuing the people of the Seven Council Fires to the Missouri River, engaging in several fights along the way. Most of the Dakota and Lakota escaped, crossing the river ahead of the troops, but they had suffered a serious loss in the food supplies gathered for the coming winter that were destroyed by the army at Big Mound. The soldiers trudged back to Fort Abercrombie on the Red River on August 1, eventually returning to Fort Snelling before winter arrived, having marched more than six hundred miles.[40]

General Alfred Sully's brigade of cavalry and artillery started from Council Bluffs in early summer but found their going much slower than Sibley's troops had. Conditions were dry, and the Missouri River low, making resupply for men and horses difficult. Sully missed his rendezvous with Sibley's column, but on September 3, 1863, his men made contact with a large camp of Yanktons and Yanktonais, Lakotas, and eastern Dakotas at a place called Whitestone Hill. Like the group Sibley had attacked, this was a hunting camp of people unrelated to the war in Minnesota, with many women and children. When the soldiers appeared, the Lakota sent emissaries to meet with them, since they did not consider themselves at war with the whites. Anxious to improve his standing with General Pope and true to the latter's stated intentions for the campaign, Alfred Sully ignored the peace offer and ordered his men to attack the camp. The Yankton, Yanktonai, and Lakota put up a desperate fight, but they were overwhelmed by the number of soldiers and the power of their artillery. Hundreds of women and children were massacred, only a few managing to escape in the chaos of the fighting. If all the bands of the Seven Council Fires had not considered themselves at war before, there could be no doubt now.[41]

General Pope felt satisfied with the results of the campaign, and the battles were generally presented to the American public as triumphs over the "savages." Indians had been killed and their camps scattered. Little Crow, viewed by many whites as the diabolical mastermind of the uprising, had been killed by a farmer in a chance encounter near Hutchinson. The "outbreak" seemed to be at an end. But if Pope's objective had truly been to end the fighting once and for all, he had failed miserably. The US–Dakota War had initially been confined to a few months of fighting over a limited region of Minnesota. But Pope's expeditions had now spread the conflict across the Upper Missouri Valley from the Red River to the Yellowstone. All of the Lakota and Dakota people had been pulled into the fight, which would eventually involve virtually every tribe in the region. The war that had begun in a Minnesota farmyard in 1862 was thus inflamed by the US military into the Plains Indian Wars, which would rage off and on for more than a decade across the Northern Great Plains to the Rocky Mountains.[42]

The following summer, a third expedition commanded by General Sully was undertaken. Troops from Minnesota included the 8th Minnesota Infantry (mounted on Canadian ponies), six companies of the 2nd Minnesota Cavalry, and two sections of the 3rd Minnesota Light Artillery. All of these troops were equipped with horses and supplies from Fort Snelling. This Minnesota brigade departed from Fort Ridgely on June 1 to rendezvous with a brigade of troops from Iowa, Wisconsin, and Dakota Territory, which was advancing up the Missouri River from Sioux City, Iowa. Its primary objective was to secure routes to the gold fields of western Montana and Idaho, establishing a number of military posts along the Missouri and several of its tributaries. These forts were also intended to take control of the region between the Red River and the Missouri to prepare for the opening of the area

for colonization, when treaties could be signed. As they had the year before, the troops encountered a large hunting camp, this time made up mostly of Teton bands, at a place on the edge of the Little Missouri Badlands known as Killdeer Mountain (Tahkahokuty Mountain). The ensuing battle was inconclusive, with most of the Lakota escaping but losing many valuable supplies. Sully continued to march through the Badlands to the junction of the Yellowstone and Missouri Rivers, harassed by bands of Lakota along his route. The expedition returned to its bases during August and September, with the Minnesota brigade reaching Fort Ridgely on October 8 after a march of 1,625 miles. Though they did succeed in establishing a series of posts, the route to the gold fields remained vulnerable and contested for years to come.[43]

A final episode in Fort Snelling's involvement with the Dakota War centered on the trial and execution of Shakopee (The Six or Śakpe) and Medicine Bottle (Wakaŋ Ożaŋżaŋ). These two Dakota leaders had gone with Little Crow into Dakota Territory after the battle of Wood Lake, and from there had escaped into Canada with a number of followers. Late in 1863, a battalion of cavalry commanded by Major Edwin A. C. Hatch, a former trader and Indian agent of dubious repute, was sent to Pembina to protect the border from this small band, which included women and children. It was a very harsh winter, and in December, Hatch, with the help of the British authorities, was able to persuade about a hundred of the Dakota to surrender and be sent to Fort Snelling to eventually join their relatives at Crow Creek. But he was disappointed that the two Dakota leaders were not among them. In a blatantly illegal scheme, Hatch bribed several Canadian officials to help his men drug and kidnap Shakopee and Medicine Bottle. The two were thus captured in Canada in January 1864 and sent to Fort Snelling the following spring.[44]

After a long delay, the Dakota leaders were tried in a proceeding similar to the trials that had condemned Dakota warriors in 1862. Held before a military court, the proceedings were longer and more formal than the 1862 tribunals, but most of the evidence against the men was circumstantial, and though Shakopee and Medicine Bottle had representation, they called no witnesses and did not cross-examine those of the prosecution. Both men were convicted and sentenced to hang. Even local papers like the *St. Paul Pioneer* were uncomfortable with the verdict, noting, "It would have been more credible if some tangible evidence of their guilt had been obtained." Despite last-minute calls for a reprieve, both were executed by hanging. The scaffold stood just outside the walls of the fort, near where the Ojibwe had executed the four Dakota men in 1827.[45]

An unexpected consequence of Fort Snelling's new role and growing demand for labor was a dramatic increase in the number of free African Americans living in Minnesota. The 1860 census for the state listed a very small Black population of 259, mostly in St. Paul. But the demand for soldiers and disruptions produced by the Dakota War created a severe labor shortage in the region. As the army began planning its campaigns beyond the Red River, it needed teamsters to drive the expedition's multitude of supply wagons. The work was hard and potentially dangerous. There were few takers among the many immigrants in Minnesota, who had more attractive options.[46]

However, the disruption of the economy in Missouri, much of which had been based on slavery, created a ready supply of able and willing workers, if the army chose to use them. The guerrilla conflict in that state was chaotic, relentless, bloody, and disruptive, with battles fought from 1861 through 1865. There had been a significant enslaved population in Missouri before the war,

and the turmoil of the struggle offered many of them the chance to free themselves. A great number made their way to St. Louis, an area under Union control. The precedent of hiring these workers as "contraband of war" had already been established in the eastern theater of war, so the quartermaster department of the Department of the Northwest quickly did the same, as reported in a *Chicago Tribune* article copied in the *Stillwater Messenger* on May 12, 1863: "The St. Louis Republican of the 7th says that the Rv. Mr. Sawyer, the special agent of contrabands in Missouri, came down from Jefferson City the day previous, with three full car loads of contrabands, collected at that place and others along the road. There are nearly two hundred in all, consisting of men, women and children—the men largely predominating. . . . We understand their destination to be St. Paul, Minn., and that they will be employed as teamsters in Gen. Pope's expedition on the plains."[47]

Even before Sawyer's teamsters arrived, however, a raft carrying families who had escaped slavery in Missouri, led by Robert T. Hickman, was towed to St. Paul by a passing steamboat. The group's arrival created consternation among St. Paulites, who had understood that only men would be brought north. Under the headline "What Will He Do With Them?" the *St. Paul Press*, a Republican newspaper, expressed a mix of racial and ethnic prejudice fairly typical of the day.

> Gen. Sibley sometime since made arrangements to have some contrabands sent up from St. Louis for teamsters. The first installment made their appearance on a barge that was brought up by the *Northerner* yesterday, and consisted of forty men, ten women and twenty-six children. This was rather more than was bargained for, and the question arises, what will be done with them? Women and pickaninnys will not render material assistance in driving mule teams over the plains, and they would probably show very large whites of

their eyes on such an occasion. We presume they will be left to garrison the Fort, while the head of the family goes roaming among the mules.

The police were very much alarmed at the appearance of such a thunder cloud, and thinking they were to be landed here, proposed to prevent it, on the ground that they were paupers. The Irish on the levee were considerably excited, and admitted by their actions that the negro is their rival. . . . On finding they were bound for the Fort, they resumed their whiskey and punches with great equanimity.[48]

And the irony was not lost to some that the *Northerner*, after landing its cargo of "contrabands and mules" at Fort Snelling, took aboard the last of the Dakota from the "encampment." In a bitterly racist note, the *St. Paul Pioneer*, the state's main Democratic newspaper—strongly anti-Black and anti-Indian, and often critical of Lincoln—sneered, "The *Northerner* brought up a cargo of 125 niggers and 150 mules on Government account. It takes back some eight or nine hundred Indians. We doubt very much whether we benefit by the exchange. If we had our choice we would send both niggers and Indians to Massachusetts and keep the mules here."[49]

In little more than a week, when a second boat arrived, the *Press* would move past its surprise and adopt a more accepting tone toward the contrabands. "The *Davenport* brought up from St. Louis last night two hundred and eighteen more contrabands, one hundred of whom were women and children. . . . Frequent applications were made on the route up to secure laborers and servants. . . . Twenty-nine were allowed to accept offers that were made. The demand for labor is so great that we understand an effort will be made to secure others in St. Louis."

But as before, these pilgrims received a rude welcome at the St. Paul levee. "The low Irish swarmed on the boat as she lay at the levee and endeavored to frighten the negroes with harum

scarum stories of the treatment they were about to receive, cursed them for not staying South, &c. . . . We conversed with several of the contrabands. . . . Most of them had been slaves in Missouri, though a few were free. The *Davenport* had on board sixty-five white teamsters and one hundred and fifty mules, and as they were all destined for the Fort, she ran up there and discharged. She will load at that point with Indians for her return trip."

And in reply to the *Pioneer*'s bitter complaint, the writer for the *Press* concluded, "We have not learned whether one hundred and fifty mules will afford the *Pioneer* society enough to drown its grief over the influx of two hundred and eighteen contrabands."[50]

Accounts and numbers vary, but it seems that at least three to four hundred African Americans made the journey from Missouri to St. Paul during these years. While some were housed at Fort Snelling, others moved to St. Paul, including the followers of the remarkable Robert Hickman; they settled in the city, forming the nucleus of its postwar Black community. Yet a significant number, at least ninety according to the post returns, worked as teamsters on Sibley's 1863 expedition. When the army began recruiting African American regiments, many of these men enlisted in the 1st Iowa Regiment of African Descent and later the 68th US Colored Infantry and served in the South through the early years of Reconstruction.[51]

⁓

Anyone who had been stationed at Fort Snelling in earlier decades might have found the number of soldiers at the post during the Civil War impressive, and they would probably have welcomed the proximity of "civilized" towns. But the routine of military life would have seemed very familiar: roll calls, drill, meals, inspections, guard mount, and other duties. Another aspect of life at Fort Snelling was also little changed: the role of alcohol in soldiers' lives. Although most of the Civil War-era

volunteers would have been considered of a more reputable character than regular army soldiers of past decades, many young American males of the era consumed large quantities of booze.

There were even more opportunities to find drink in and around 1860s Fort Snelling than there had been in earlier decades. The sutler's store, owned and operated as in the past by General Sibley's brother-in-law, Franklin Steele, sold alcohol within the fort. Steele always claimed it was in the form of less intoxicating beer rather than whiskey, but his critics suggested otherwise. The Faribault House in Mendota was another popular spot. Operated as a hotel by members of the family, it featured a billiard room and bar on the main floor. The place was a constant source of trouble, drunkenness, fights, and even a few killings. If a soldier could obtain a pass, it was easy enough to travel to any of the hotels or taverns in St. Paul, or the neighborhood of Minnehaha Falls, to find all kinds of entertainment. Or the more adventurous could simply climb over the wall and take his chances among the shanties located in the woods nearby, often run by an enterprising Irishman selling shots from a jug or two of rotgut whiskey.[52]

As ever, drunkenness and other infractions continued to result in many court-martial trials and punishments. With so many troops at the post, sometimes over a thousand, the military jail was at times packed with prisoners. In 1864, General Sibley ordered the building of a new military prison for the District of Minnesota. Though it faced bureaucratic delays, the stone structure was eventually completed in 1865; in later years, it served as the post exchange.

As it became apparent the war would not be over quickly and the first enthusiasms of volunteer enlistments faded, Congress passed the Militia Act of 1862, empowering the president to draft up to 300,000 men for nine months of service. For a while, the threat of a draft kept enlistment levels high enough to avoid forced conscription. By early 1863, still more manpower was needed for the Union Army. Congress passed the Enrollment Act on March 3, 1863, requiring all males between the ages of twenty and forty-five to register for the draft. Though the draft did not take place in Minnesota until 1864, the influx of draftees required new construction at Fort Snelling. A draft rendezvous—a set of barracks and kitchens surrounded by a stockade—was built on the prairie just north of the round tower.[53]

Drafted men were housed in this "pen," where they were processed, organized into temporary companies, and issued clothing before being sent to their assigned regiments. Their length of stay depended largely on the availability of transportation. To watch over the draftees and the enrollment and conscription process, Companies C and K of the 23rd Regiment of the Veteran Reserve

Corps, about 120 men, were assigned to Fort Snelling. These were disabled soldiers serving out the remainder of their enlistment in the VRC, or disabled soldiers whose enlistment was over but who wished to perform further service. Their service in less strenuous guard duty freed regular troops for field service.[54]

Following the end of the war, Fort Snelling was the site of mustering out of service for most Minnesota soldiers. A few units were hailed by joyous parades, but most quietly returned to the fort, waited out the days while their paperwork was completed, received their pay, mustered out, and went home. For some posted in the West, or who enlisted late in the war, the wait went on into 1866. After April 1865, the numbers assigned to the fort gradually diminished. In December 1865, two companies of the 10th US Infantry arrived, and from that time on, the "regulars" gradually replaced the volunteers. During the following year, the post continued to gradually wind down some of its activities with the closure of the draft rendezvous and the relief and discharge of the Veteran Reserve troops. Apart from the discharge of the remaining Minnesota volunteer units, the most active contingent at the fort seemed to be the quartermaster department, which still had forty-two civilians on its payroll, including twenty-two teamsters, four wagon masters, three blacksmiths, and five clerks.[55]

Thousands of young men had passed through Fort Snelling, and hundreds of them never returned. The war in the South and the war in Minnesota forever altered the state. What had remained of the old fur trade culture had been largely swept away and seemingly the Dakota people with it. The Dakota homeland was now becoming a milling, transportation, and agricultural center, just as many Americans had predicted and desired.

The Dakota at Crow Creek would continue to suffer; hundreds of people, mostly children, died of malnutrition and disease in the first three years. In 1866, President Andrew Johnson ordered the release of the men who had survived imprisonment at Davenport, and they were reunited with their families. Many moved to other reservations in South Dakota and Nebraska.[56]

But even as thousands of immigrants poured into the region, Dakota people remained. Some hundreds of Dakota people lived at Mendota and Faribault after 1862, under the protection of Henry Sibley, Alexander Faribault, and Episcopal Bishop Henry Whipple. Others returned to the state in the years that followed, eventually establishing communities at Prairie Island, Shakopee, Lower Sioux, and Upper Sioux. They continued to visit Bdote, Mini Sni, and other sacred sites—and continued to show that Minnesota is a Dakota place.[57]

Fort Snelling from Pike Island, about 1890, showing the
Minnesota Valley Railroad's trestle. Later part of the Chicago and Northwestern system, the line followed the south
shore of the Mississippi to Mendota, then traveled up the
Minnesota River; a spur cut across the Minnesota River
below the fort and continued to Minneapolis via Hiawatha
Avenue. The right-of-way is now a paved hiking and biking
trail. *Watercolor by Michael J. Miller, MNHS.*

10. *Expansion and a New Role*

SOON DISCUSSIONS ON THE FATE OF FORT SNEL-
ling began once more. Quartermaster General
Montgomery Meigs wanted to use the fort and
most of its buildings as warehouse space for supplies being forwarded to western outposts. In a
letter to Secretary of War Edwin Stanton, written in June 1866, he argued that the fort and one
square mile of the reserve should be retained by
the government.

> In addition to the Post buildings a Draft Rendez-
> vous has been built since the war and they are
> all within the square mile recommended to be
> retained. They are the best buildings of the kind I
> have ever seen. These I would fill with such surplus
> stores as we now have in buildings at St. Louis and
> Cincinnati for which we are paying rent. In this
> way we can store such surplus property as we want
> to keep for future use, at no expense for rents. . . .
> This will be a good place to store them, as the Posts
> of Ripley, Ridgely, Abercrombie and Wadsworth
> will probably be kept up for some years yet and
> this is the proper place to supply them from.[1]

The governor of Minnesota, Stephen Miller,
also expressed an interest in the buildings and
property within the military reserve for use as a

home for disabled soldiers. He wrote Stanton on
the subject in July and Alexander Ramsey as well,
noting that the army had made similar provisions
for other states. For the time being, however,
the army saw the fort's role as a location where
troops could be concentrated and a rendezvous
site in the event reserves were needed, and these
requests were turned down.[2]

Local desires confronted national priorities.
The end of the Civil War left the North with a
burgeoning economy and rapidly expanding population. In 1866, anticipating further American
expansion in the West, the army reorganized itself
into four divisions: the Military Division of the
Atlantic, the Military Division of the South, the
Military Division of the Missouri, and the Military Division of the Pacific. The Department of
Dakota, including Minnesota and Fort Snelling,
was part of the Division of the Missouri, which
was commanded by General Philip Sheridan and
headquartered in Chicago.

With southern obstruction removed from
Congress, the way was open for the rapid creation
of new states on lands taken from Indigenous
peoples and for federal spending on internal

US Army Model 1872 bison hide overshoes worn by Dr. James B. Ferguson, acting assistant surgeon from 1870 through 1891. He served at Fort Totten in 1875 and Fort Yates in 1877. *MNHS.*

improvement, particularly railroads. Minnesotans were finally connected by rail to the East in 1867. The McGregor Western Railway, building track northward from Iowa, linked up with the Minnesota Central Railway, coming south from Minneapolis, creating a route via Austin to Prairie du Chien, Madison, Milwaukee, and Chicago. This route was expanded the following year, and other routes, particularly the St. Paul and Pacific and the Northern Pacific, were reaching into North Dakota by 1871.[3]

The new Department of Dakota was headquartered in St. Paul under the command of General Alfred H. Terry. The city was seen as a more convenient spot for communication and for the disbursal of supplies to frontier posts. Most supplies still arrived by boat, and it was easier and cheaper to store goods temporarily in town for shipment by rail to St. Cloud, where they could be sent by wagon and eventually by rail to posts farther west.[4]

These and other factors once more brought the future of Fort Snelling into some doubt. Franklin Steele had never made full payment of the sale price, but he retained his interest in the location and had his 1857 purchase agreement in hand.

He sued for return of the property and the "rent" that he claimed the army owed for using it. The controversy over the earlier sale, particularly the fact that it had involved Secretary of War Floyd, who had served as a Confederate general, made the army and Congress reluctant to accept Steele's terms.

In 1867, William Tecumseh Sherman, now a lieutenant general commanding the US troops in the West, made an inspection of the western posts, including Fort Snelling. Soon after, he wrote the War Department a letter that would heavily influence its attitude toward the fort for the next thirty years. He stated unequivocally that the site of Fort Snelling had great strategic value and "should be held by the United States forever . . . should the site now pass into private hands, it would have to be repurchased at some future time at a vast cost." He agreed the earlier sale had been a fraud, but he believed the fault lay with former Secretary of War Floyd, and not Franklin Steele. Sherman urged that the War Department now settle with Steele.

It took another three years before the War Department came to an agreement. A board of officers estimated that Steele owed the government $68,200 from the original sale and set the value of the fort and other buildings at $12,920. They further allowed that $17,250 was owed Steele for rent due from nine years and seven months of army occupation of the site at a rate of $150 per month. This left Steele owing $38,030, for which the government retained 1,521 acres of land valued at $25 per acre. The remainder of the 1857 reservation lands, some 6,395 acres, was deeded to Franklin Steele. The value of these lands in 1870 far exceeded its already undervalued 1857 price, and, as General Sherman predicted, the government would end up buying back land for the expansion of Fort Snelling.[5]

Much of the ongoing conflict with Native peoples—particularly the Lakota, Cheyenne, and other Indigenous nations living on the Great Plains—would take place in the Department of Dakota and the Department of the Platte. For the remainder of the 1860s, the army had tried to continue what Sully's troops had started in 1864 by maintaining a route to the gold mines of Montana and Idaho. This had proven difficult. The Lakota and their Cheyenne allies, having learned from earlier fights, no longer took on massed troop formations in open battle. Using their mobility and taking advantage of the terrain, they inflicted severe losses on the overextended troops.[6]

The army had changed as well. By 1868, the volunteers and draftees were gone, and the enlisted men were once again mostly immigrants, predominantly Irish, with a sprinkling of former Confederate soldiers known as "Galvanized Yankees," for whom soldiering was the only career they knew. Many of the younger officers were West Point graduates, trained to fight in organized wars against nations like their own with little knowledge of Indigenous people or their style of fighting. Even the more senior officers like Generals Terry, Sheridan, and Sherman had gained most of their experience and fame fighting in the Civil War. The army in the West acted much like a police force, patrolling territory between isolated posts and occasionally reacting to raids or attacks. As in past decades, there were insufficient numbers to effectively control the vast areas involved. But the importance of mounted troops was recognized far more than in the pre–Civil War army.[7]

In 1872, the troops under the command of the Department of Dakota included five regiments of infantry and two more of cavalry; in 1873, another regiment of cavalry, the 7th, was added and assigned to guarding the Northern Pacific Railroad survey. Most of the regiments were spread out in battalion- or company-sized detachments at the many outposts manned by the army across Minnesota, Dakota Territory, and Montana Territory. These distant posts were supplied and supported through the department headquarters in St. Paul. Detachments of troops and recruits traveled through St. Paul with its river and rail connections on their way to western outposts. Fort Snelling served as a strategic location where troops could be billeted, reinforced, and reequipped before heading out to bases farther west.

The army was enforcing a government policy aimed at "civilizing" the Native American nations living on the Great Plains by turning them away from their traditional lives and making them into farmers, much as had been attempted in Minnesota and elsewhere. Yet Lakota resistance to white incursions along the Bozeman Trail in western Montana had been so intense and costly to the army that the government had abandoned the area. In the Fort Laramie Treaty of 1868, signed with the Oglala, Miniconjou, and Brulé bands of Lakota, the Yanktonai, and the Arapaho, the government agreed to close the Bozeman Trail, established a reservation for the Sioux that encompassed all of what is now South Dakota west of the Missouri River (including the Black Hills, a sacred area), and further recognized a large area north of the North Platte and east of the Bighorn Mountains as unceded territory on which whites were not permitted to settle and could not cross without permission from the tribes. The Indigenous nations agreed to stop opposing railroads; they were promised various annuities and tools for farming, and they agreed to allow travel along certain specific routes.

Even from the start, this treaty proved unenforceable, and army commanders, particularly Sherman and Sheridan, were skeptical that they could prevent whites from intruding on Indian lands. They also believed only force could compel the Lakota, Cheyenne, and Arapaho people to accept reservation life. The treaty also had the effect of dividing the Lakota. Respected leaders such as Red Cloud and Spotted Tail and their followers, feeling that warfare with the whites could not go on indefinitely, wanted to try living on the agencies assigned to them. Others were determined to maintain their traditional ways on the unceded lands. These people, including some Northern Cheyenne and Arapaho, coalesced around a charismatic and successful young warrior named Sitting Bull.[8]

———

The peace began unraveling in 1874. The United States was in a deep depression; rumors of gold deposits in the Black Hills put pressure on the government to allow a survey of the region. In the summer of that year, the 7th Cavalry Regiment, led by Lieutenant Colonel George Armstrong Custer, set out from Bismarck, North Dakota, to find a suitable location for a fort—and to verify the rumors. Civilian members of the Black Hills Expedition confirmed them, and excited newspaper accounts spread the word. Prospectors stampeded into the Black Hills. This blatant violation of the Laramie treaty angered both the Lakota and the Cheyenne, for whom the Black Hills held spiritual and economic importance.

Fighting resumed between the traditionalists, led by Sitting Bull, and the US Army. Again, the army in the West received little in the way of appropriations or numbers of troops to support their efforts; much of the army was still occupying the former states of the Confederacy. The Lakota and their allies living off the reservations in the unceded territory fought with intelligence and courage to defend their way of life. Their numbers included famed war leaders like Crazy Horse and Gall, and enough buffalo remained on the plains to sustain them. Indeed, they were living better than those who had moved to the reservation. There were even reports of young warriors and their families leaving the agencies to join Sitting Bull.[9]

In the summer of 1876, the army made a determined effort to encircle and trap Sitting Bull and his allies, deploying three columns of troops: one under General George Crook came north from Fort Fetterman, Wyoming, in the Department of the Platte; another under General John Gibbon traveled east from Fort Ellis, Montana; and a third led by General Alfred Terry, Commander of the Department of Dakota in charge of the overall operation, moved west from Fort Lincoln. The troops in Terry's command, some of whom had previously been at Fort Snelling, included the 7th Cavalry, led by Lieutenant Colonel Custer. The columns were to converge somewhere along the Bighorn River in south-central Montana. But Crook's troops were attacked by Crazy Horse on the Rosebud River and compelled to turn back. Then Custer, contrary to instructions, moved his cavalry far past the support of General Terry's mixed infantry and cavalry. He discovered the Lakota, Northern Cheyenne, and Arapaho camped along the Little Bighorn River and rashly decided to attack. The result was a disastrous defeat. On June 25, Custer and five companies of his 7th Cavalry, more than 250 troopers, were wiped out by the large force of Indigenous warriors. Terry's troops rescued the remnants of the 7th Cavalry and reported the disaster to the army.[10]

The majority of Americans learned of the 7th Cavalry's fate in early July, as the nation was celebrating its centennial. It seemed to most citizens incredible and outrageous that a group of "savages" could so thoroughly defeat the troops, especially when commanded by such a well-known and popular officer. Public opinion now demanded that Custer be avenged. Divisions over Indian policy or desire for more humane treatment were swept aside. Little Bighorn was a humiliating defeat for the US military, but the public response was to give the army everything it desired to suppress the Lakota and their allies.

Thus, Custer proved more useful to the army as a dead hero than a living commander. Within a few weeks, reinforcements were sent from other departments to the Department of Dakota, and Congress gave the army control over all Indian reservations and agencies. New forts were authorized in areas of the Black Hills and western Montana that had been designated Indian lands in the Fort Laramie Treaty. By the end of the following winter, the followers of Sitting Bull had either fled to Canada or accepted a place on the reservations. The Battle of the Little Bighorn had been a disaster for them as well.[11]

Though the Department of Dakota included a vast territory and thousands of troops stationed in outposts as far away as Montana, the garrison of Fort Snelling remained relatively small, and only minor additions were made to the post to accommodate the few companies of the 10th Infantry stationed there; the department's offices had remained in St. Paul since 1867. This situation began to change when Congress passed an act on June 18, 1878, requiring that all department headquarters be located on military posts. General Terry thus had a problem: Fort Snelling had no suitable buildings to house his headquarters. He wrote General Sherman,

> If such be your desire and the desire of the War Department, I think that the members of Congress from Minnesota would interest themselves very strongly to obtain a special appropriation for the purpose. They know the views which you expressed a few years ago upon the subject of retaining the reservation at Snelling as a place of arms for the Northwest, and they think that their local interests would be subserved by the establishment of the headquarters there. Of course, they would not think of moving for an appropriation without the approbation, or at least the assent, of the Secretary. With it, I have no doubt that they can and will obtain a reasonable sum.[12]

Sherman soon confirmed the desire that the headquarters go to the fort and that he "wish him [Terry] to submit plans and estimates for good buildings on the Snelling Reservation."

A few weeks later, on December 27, 1878, the chief quartermaster for the Department of Dakota submitted a list of buildings and their estimated costs, calling for no fewer than thirty-seven buildings—probably the greatest expansion of the post since it was built. The plan also gives some insight into the size of the bureaucracy, both military and civilian, required to run such a large department, and the huge quantity of space rented in St. Paul. It was no wonder General Terry thought the locals would feel their "interests had been subserved"—construction would bring much wealth to the community.

The proposed expansion of Fort Snelling in 1878 included thirty-seven new buildings:

One building, two stories with basement, cistern and furnace complete, for offices of department headquarters, with storerooms, &c.	$14,800
One building, quarter for commanding general	8,250
Fourteen buildings, quarters for officers on the department staff, including for field officers, $4,870 each	68,150
Ten buildings, quarters for necessary general-service and enlisted men and civilian employees connected with department headquarters	25,000
Ten buildings, stable accommodations for regulation allowance of horses for officers, and a forage-house	4,400
One building, mess-hall, kitchen, $c, for employees	2,100
Fencing	2,270
TOTAL	125,000[i]

As he also anticipated, the quartermaster department had no resources to foot such a bill, so a special appropriation was required. Secretary of War George W. McCrary approved the request. Congress appropriated $100,000 in March 1879 for the building of the headquarters and some of the related structures, including quarters for the commanding officer and staff officers. The following year, the department requested an additional $100,000 to complete the work. In his descriptive request for the additional funds, Department Quartermaster Charles H. Tompkins showed how the fort was becoming an icon of the state's history, a powerful reminder of the colonial past.

> The site selected for the headquarters buildings, the associations and surroundings of the neighborhood, combine to invest Fort Snelling with a peculiar charm. This old post, the pride and strength of a generation of pioneers in this country, is midway between the cities of St. Paul and Minneapolis, and is a central point of interest to residents of both as well as to all tourists in the country. Thousands of people visit the post and falls of Minnehaha and Bridal Veil. . . . I believe it will enhance the value of the many attractions of the place and heighten the interest felt by the people of Minnesota, if the establishment of department headquarters, within sight of the famous old post, is made permanent, handsome, and complete in all details of structures and landscape culture.[13]

In his endorsement of the report to the secretary of war, General Sherman again stated his opinion about Fort Snelling: "I regard it as a strategic point which should always be held by the United States, and am therefore disposed to recommend almost any outlay which will make it valuable as a permanent military site."[14]

The secretary of war to whom this proposal was put embraced it with open arms, which was no surprise. He was none other than Alexander Ramsey, who succeeded McCrary in December 1879. After two terms in the US Senate, the former governor was well positioned to advocate for this and other expenditures to improve the fort.

Almost simultaneously, another funding proposal was making its way through Congress. The idea of building a bridge across the Mississippi River at Fort Snelling actually predated the move of the Department of Dakota headquarters. In May 1878, Secretary of War McCrary presented to the Senate a detailed report, including surveys of possible bridge locations and a proposal from the St. Paul Chamber of Commerce, stating, "The city of Saint Paul is interested in erecting a free bridge at this point and would be willing to join the government and share a reasonable portion of the expense. . . . The United States own a large tract of land at and near the fort which would be largely enhanced in value by the erecting of a bridge."

General Terry endorsed the proposal, echoing these points and noting the difficulty of road communication with St. Paul. The ferry crossing was always slow and sometimes delayed by changing river conditions. He pointed out that the site recommended for the bridge would bring the associated road near the north wall of the old fort, but that the inconvenience would be "so slight that I think that it ought not to weigh against the interest of the large population to which the bridge is of vital importance."[15]

While all was enthusiasm for the bridge from the St. Paul side, some in Minneapolis, its neighbor and growing rival, saw the project as "an effort to draw away from Minneapolis some of the trade of Richfield, Carver, Shakopee, Chaska, and kindred territory." There was probably more than a little truth to this statement. The *St. Paul Daily Globe*, in a long article advocating for the bridge, described the greatest advantages as avoiding the "dangerous Snelling hill [the steep road leading

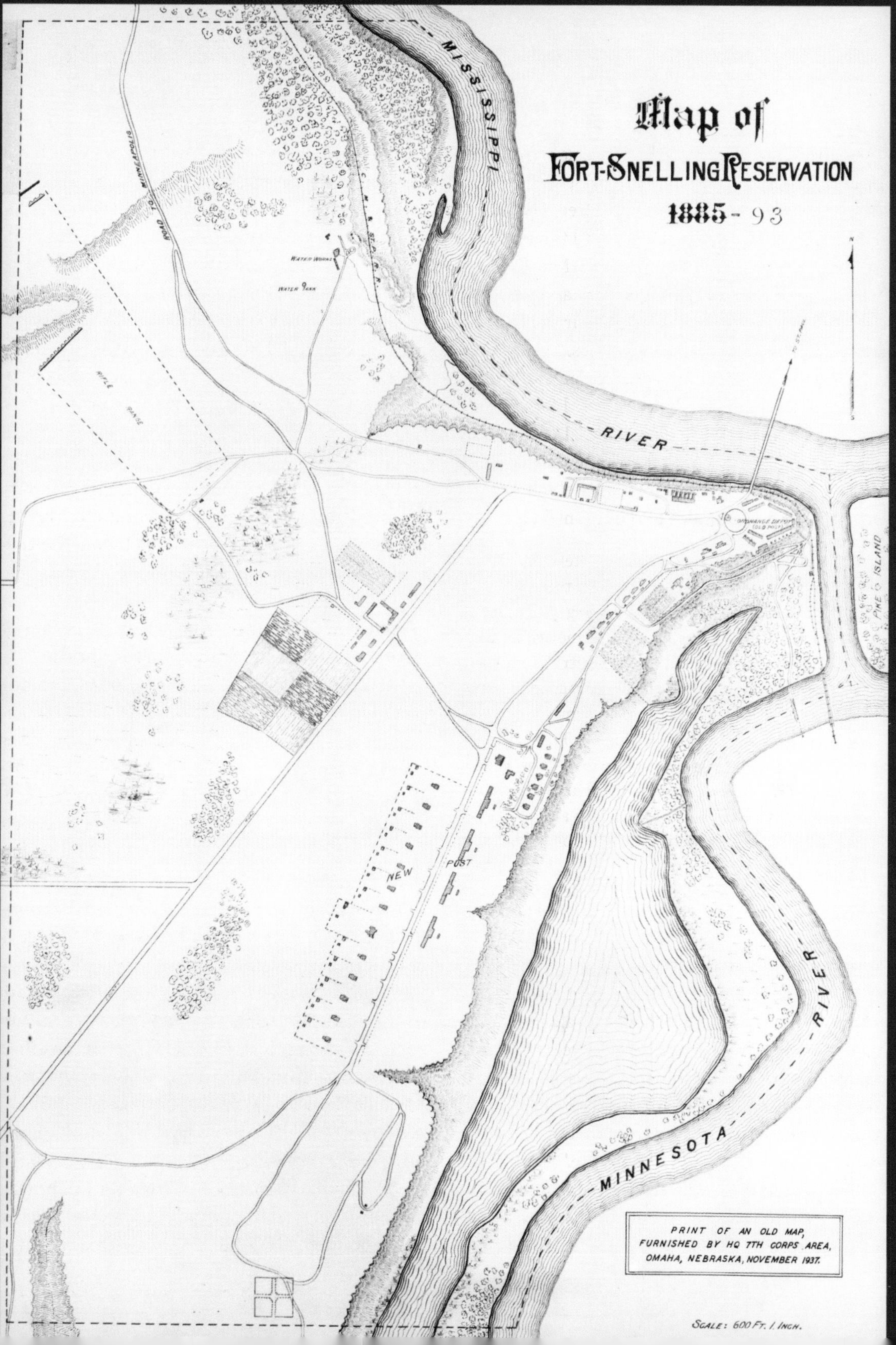

Map of
FORT-SNELLING RESERVATION
1885-93

MISSISSIPPI

RIVER

WATER WORKS
WATER TANK

RIFLE RANGE

ORDNANCE DEPOT

PIKE ISLAND

NEW POST

MINNESOTA RIVER

PRINT OF AN OLD MAP,
FURNISHED BY HQ 7TH CORPS AREA,
OMAHA, NEBRASKA, NOVEMBER 1937.

SCALE: 600 FT. 1 INCH.

The Fort Snelling bridge, between 1905 and 1909. At the south end of the bridge, travelers passed close by the north wall of the fort and the round tower. The barracks were torn down in 1905. *Photo by Francis L. Wright, MNHS.*

up to the fort from the river landing] . . . and the delaying, risky, tax-gathering ferry—two obstacles in the line of intercommunication, so serious as to have proven a positive blockade for years . . . between a large and rapidly growing agricultural region and its natural outlet at St. Paul."[16]

The construction went forward with surprising speed. The estimated cost of the structure was $165,000. Congress appropriated $100,000 in early 1878, and Ramsey County passed a bonding referendum for the remaining $65,000. In early December, the *Globe* proudly reported the near completion of the Fort Snelling bridge under the headline, "St. Paul Takes Another Stride for Business Supremacy." The considerable ironwork construction of the deck-style bridge, with the iron trestles below the roadway, was done by the H. E. Horton Company of Rochester, Minnesota. The iron itself was manufactured by the Carnegie Company and fabricated by the Keystone Bridge Company of Pittsburgh. It was a structure of "magnificence and massiveness, combined with strength." A few months later, the St. Paul Bridge Commission concluded its duties by passing a resolution stating, "We have fully completed said bridge, it is in all respects finished open to travel and irrevocably dedicated to the public and free to the United States of America and all the people thereof."[17]

Even as the bridge was being built, contracts were let for the new military headquarters and related buildings. The headquarters building, commandant's quarters, and quarters for field officers and captains were all designed to be modern buildings with gas, central heat, and other conveniences. New waterworks built at Coldwater Spring—a springhouse, reservoir, pump house, and water tank—provided a pressurized source of water.[18]

It was, all in all, a remarkable change. In a few years, a new Upper Post of modern buildings rose outside the walls of Old Fort Snelling, and the new post enjoyed much-improved connections to St. Paul, which itself was now more closely joined to the rest of the nation by direct rail connections. No longer would winter mean isolation. As the railroads moved north and west, the city became a transportation hub. To the north of the fort, Minneapolis, with its waterpower and its own rail connections, was rapidly growing to challenge the capital city in population and wealth.

Starting in 1882, the headquarters of the Department of Dakota was located in its new building on the post.

Compared to the troops serving in the western outposts, those stationed at Fort Snelling had comparatively luxurious duty, with regular rations and pay and towns close by. From 1872 through 1878, one or more companies of the 20th Infantry, along with its headquarters and band, were located at Fort Snelling. In 1878, these were replaced by six companies of the 7th Infantry along with its headquarters under the command of Colonel John Gibbon. Gibbon was one of the best-known and -respected officers in the army

The springhouse and reservoir at Coldwater Spring, 1885. *MNHS.*

The headquarters of the Department of Dakota, 1880s–90s. *MNHS.*

of that day, having served with distinction in the Civil War as commander of the Iron Brigade and the 2nd Division of the 2nd Corps of the Army of the Potomac, a unit that included the 1st Minnesota Infantry. When they arrived at Fort Snelling, the 7th Infantry had just concluded a long and bitter campaign against the Lakota, Cheyenne, and Nez Perce in Montana and Idaho. Gibbon, referred to as "General," his Civil War rank, and his troops, particularly the band, were a frequent presence in St. Paul.

In 1883, Companies B, C, I, and F of the 25th Regiment, plus the headquarters and band—474 men in all—were assigned to Fort Snelling. This was one of the US Army's four segregated, all-Black regiments, which also included the 9th Cavalry, 10th Cavalry, and 24th Infantry. Collectively, they were known as "Buffalo Soldiers," a name thought to have been given to the soldiers

of the 10th Cavalry by either the Cheyenne or the Comanche, who occasionally fought them. The term, which was possibly derived from the soldiers' curly hair, their bravery in combat, or even the buffalo robes they used in winter, was applied to all the segregated units. Apparently, the soldiers did not use this term in referring to themselves; it was picked up and repeated by white journalists.[19]

The 25th Regiment was created by an act of Congress on March 3, 1869, which consolidated two earlier US Colored Troop regiments, the 39th and 40th, to form the new unit. Prior to their arrival at Fort Snelling, the 25th had served in the Southwest, escorting government contractors and guarding railroad construction crews in Texas. In the course of these assignments, the troops of the 25th had several clashes with Comanche Indians and outlaws along the border with Mexico. Besides the troops at Fort Snelling, Companies A,

Company B of the 25th Regiment at Fort Snelling, 1883. *MNHS.*

D, H, and K of the regiment went to Fort Meade, and Companies E and G went to Fort Sisseton in Dakota Territory.

The historian for the regiment remarked of their stay at Fort Snelling, "The next five years, 1883 to 1887, were destined to be the least eventful in the regiment's history, and the regimental returns for this period contain, month after month, under the 'Record of Events' but one monotonous entry—'no change.'"[20]

From the beginning of their stay, according to newspaper coverage, the regiment's soldiers, and particularly the regimental band, seem to have impressed numerous citizens of the state, and they appeared at many ceremonies and events. Early in the regiment's career at Fort Snelling, however, a local citizen took offense at the presence of African American soldiers at the front of a parade in St. Paul. The occasion was Decoration Day, now Memorial Day, in 1883. Signing himself "Clay" (presumably for Henry Clay, a former senator), he wrote to the *Daily Globe*:

> What a sight for a white man to look at, be he Republican or Democrat, to see a procession made up in honor of our departed heroes with the whole head and front made up with the colored musicians and soldiers. . . . I believe in giving the colored people everything that is their due, but I certainly do not believe in our own race who have fought to free themselves from the yoke of czars and kings and who have given up life and friends to liberate an ignorant class of people from slavery to quietly submit to such an outrage as was shown in to-day's procession. . . . Such proceedings will lead to a more serious trouble than slavery did, for true white men will not always quietly submit to be placed at the tail end of the procession with the solid front made up of Black soldiers and musicians.[21]

This sort of racism was undoubtedly common among St. Paul whites, as in the rest of American society. Prejudice was deeply planted; the community's days of racial ambivalence rooted in the fur trade had long since passed away. Though the Black community of St. Paul was relatively tiny, less than two percent of the population, they had struggled to obtain a degree of equality. They had won the right to vote in state elections in 1868, two years before passage of the Fifteenth Amendment, and they had desegregated local schools. This was not a passive minority. Their success was accomplished through political organizing and the support of the local Republican Party. Yet economic opportunities for Blacks remained limited, and housing was segregated. It is easy to imagine that members of that community took pride in the men of the 25th Regiment and appreciated the repost sent by Jose Cabeon, one of the soldiers, that soon appeared in the *Globe*.

> Please allow me space in your columns to respond to Mr. "Clay's" comments (our patriotic and philanthropic hero) in the *Globe* dated May 31, 1883, relative to the "black musicians and soldiers leading the processions," and our ignorance, slavery, liberation, etc. what he would have done, had he belonged to one of those military companies. We only wish he had belonged to one of the companies, so he could have stepped out, then we would have seen his ignorance. Mr. "Clay" has mistaken the "black musicians and soldiers." They have not all been slaves and they are not all ignorant, not as much I presume, as some of Mr. Clay's beloved countrymen, as I know he must be a countryman, he can not be a resident of the city of St. Paul, among our friendly, and social citizens of that city. . . . I do not see why Mr. Clay should kick simply because he was not in command. I suppose, Mr. Editor, as far as liberation is concerned, you know, we fought: as hard for that, I suppose, as Mr. Clay, or probably harder, as I have an idea that he was in a coal shed during the war. This communication is prompted by no feelings of disrespect to any one, but in the belief that our friend Clay is not worthy of society, or he would have been in the procession.[22]

The 25th Infantry was not the only outfit in the garrison. Battery E of the 4th Artillery was also stationed at Fort Snelling. Its members shared similar duties with the 25th Infantry, and they may have been the white soldiers referred to in Clay's letter.

During the 1880s, the role of Fort Snelling and its garrison continued to evolve as part of the Department of Dakota. The fort was regularly used as a place to house recruits destined for units stationed farther west. Groups of twenty or more recruits arrived at St. Paul via steamboat or railroad and were sent to the barracks at Fort Snelling. Officers and soldiers from the garrison escorted the recruits to their assigned posts farther west. The fort was also the site of regular marksmanship competitions among the various regiments in the department. Each sent a representative to compete, and the contests frequently drew members of the public as spectators. This was part of General Alfred Terry's effort to improve marksmanship among the troops.[23]

In May 1888, the 25th Infantry was transferred to Montana, and its place was taken by one of the best-known regiments of the regular army, the 3rd Infantry. Considered the oldest regiment in the army, it had served with distinction from the Revolutionary War through the Civil War. Its arrival at Fort Snelling coincided with the War Department's decision, supported by Minnesota Congressman Edward Rice, to request an appropriation of $150,000 to build new, modern barracks for eight companies of infantry at Fort Snelling. With the end of the fighting on the Great Plains in the late 1870s, the army sought to realign its bases according to new needs. Small, isolated garrisons, difficult to staff and expensive to supply, gave way to larger posts located near transportation centers. With St. Paul's emergence as a railroad hub, Fort Snelling seemed a perfect site. Over the next decade, "under the liberal policy of the War Department and by the active interest of the members of the Minnesota delegation in congress," facilities at the post expanded to incorporate an entire regiment. The immediate impact of the 3rd's arrival, however, was to crowd out the headquarters of the Department of Dakota, which returned to a city block in downtown St. Paul, where it remained until it was dissolved in the army reorganization of 1911.[24]

Garrison duty for the 3rd Infantry was not quite the sedate, even monotonous, experience of most earlier regiments. Its colonel, Edwin Mason, frequently led the men on maneuvers outside the military reserve. In 1889, they made a "practice march," according to the *Minneapolis Tribune*. "The route of march will be west from the post across the country to Lake Minnetonka, around the lake to its south shore, thence to the Mississippi river near Anoka, and thence by Coon creek, Lake Johanna and other points to the post. Camps will be made at the Hotel St. Louis, at Zumbro Heights, at Chapman's (at the head of the lake) and at Maple Plain. The regimental band under the leadership of Prof. Win. Griffin, will give daily concerts to enliven the time."

In the summer of 1891, most of the regiment, again including the band, made a cross-country march from Fort Snelling to Camp Douglas in Wisconsin, much to the delight of civilians along the route. The *Chatfield Democrat* noted their passage through town: "They were halted at city park . . . the command is a fine body of youthful looking men, who appeared to be in excellent discipline and perfect drill." While there was certainly a public relations element to these exercises, they did have some military value, by introducing soldiers to long marches and providing a relief from garrison routine.[25]

On at least one occasion, the military utilized its mobility strategy by sending companies of the 3rd Infantry on an emergency deployment, against not some foreign invader but American workers. In April 1894, four companies were sent out from Fort Snelling by train to North Dakota to "assist" in the arrest of striking Great Northern railway

The fort became an increasingly romantic subject for Minnesota artists and a photographer. Alexis Jean Fournier, known as the last of the Barbizons, painted a long view showing the new bridge in 1885 (*oil, MNHS; gift of Mrs. E. A. Hendrickson*). Nicholas Brewer, _____, born in Minnesota in 1857, painted the hexagonal tower in the 1890s (*oil, MNHS*). **OPPOSITE** In 1889 Henry Peter Bosse, an engineer and photographer who took cyanotypes of scenes all along the Mississippi, made this image (*National Gallery of Art*).

workers. The *Minneapolis Tribune* reported they were "fully equipped . . . for a long campaign."[26]

This was the same time period as the long and bitter Pullman Strike and general labor unrest across the country. There was fear of mob rule on one side and violent repression on the other. "Thousands of men have been massed at this point," one journalist wrote of Fort Snelling. "When there were threatened Indian outbreaks and . . . with mobs in cities like Chicago or Milwaukee or Omaha, a thousand men could be on board the cars and on their way to the threatened point at any time."[27]

"Thousands" were not sent out from Fort Snelling for any such purpose, but most of the 3rd Infantry was sent out to take part in the Spanish-American War—specifically the invasion of Cuba. On April 19, 1898, six days before the formal declaration of war, they packed up their gear and traveled by train to Mobile, Alabama, then by ship to Cuba, where they served until late August. There they fought at El Caney and San Juan Hill. After ten years in the state, the regiment had recruited

plenty of Minnesotans into its ranks—593, by the end of the Spanish-American War and the fighting in the Philippines. A number of them were wounded, and several succumbed to the epidemic of yellow fever that surged through the regiment in the short time it was in Cuba.[28]

Though Fort Snelling was not involved with the formation of state volunteer regiments, it was used as a recruiting depot for the regular army, particularly the 3rd Regiment. New enlistees began their training at the post from the cadre of soldiers and officers remaining while the bulk of the garrison was in Cuba. These men would find themselves involved in a bizarre, tragic incident, known as the Battle of Sugar Point.[29]

The chain of events started in April 1898 with the arrest of an Ojibwe man, a member of the Pillager band at Leech Lake named Bagone-giizhig (sometimes spelled Bug-O-Nay-Ge-Shig) or Hole in the Day. He was not related to the more famous chiefs of the Mississippi band who bore this name, but a member of the Leech Lake band, who like others, was trying to make a living on the lands left to them. For motives not completely clear, the federal marshal, Robert T. O'Connor, charged Bagone-giizhig with bootlegging and brought him to Duluth for trial. There was little evidence against him, so he was never prosecuted. The court ordered him released, but the officials who sent Bagone-giizhig all the way to Duluth did nothing to help him get back to his village, nearly two hundred miles away. It took him weeks to walk home, and he vowed never to be arrested again.[30]

The following September, the same US marshal tried to detain him and a friend as witnesses for a trial in Duluth. Bagone-giizhig refused, and with the assistance of a group of villagers, he escaped. The marshal sent to Fort Snelling for help, and twenty soldiers from the 3rd Infantry arrived in nearby Walker on September 30. Attempts by the local Indian agent to persuade Bagone-giizhig to turn himself in proved unsuccessful. He was leery after his previous experience. The marshal and the agent also knew of other reasons for discontentment at Leech Lake: band leaders had recently sent a polite petition to the secretary of the interior, requesting an investigation of timber poaching on their tribal lands by a white-owned lumber company. No marshals or soldiers or other law enforcers had seen fit to stop the loggers' depredations.[31]

On October 4, 1898, another seventy-seven soldiers from Fort Snelling commanded by veteran Captain Melville Wilkinson arrived at Leech Lake. On the following day, the troops and civilian officials went by boat to a location known as Sugar Point on Bear Island, where Bagone-giizhig was said to be staying. After their arrival, a confrontation developed with some Ojibwe camped there. Accounts differ as to how the shooting started: some said a clumsy soldier had failed to engage his rifle's safety; Ojibwe witnesses said the soldiers began to fire at Ojibwe women in a canoe. But the soldiers, most of them recent recruits, were quickly pinned down by Ojibwe men concealed in the surrounding woods. After hours of shooting, the Ojibwe allowed

the soldiers to retreat. Some left in the boats that brought them, others were brought out the next day. Six soldiers were killed in the affair, including Captain Wilkinson, and ten more were wounded; a sentry also killed a tribal policeman whom he had mistaken for an attacker.[32]

These were the first armed hostilities between the US and the Ojibwe in eighty-nine years. For the army, it was an embarrassing episode. Wild stories telling of the massacre of dozens of soldiers, with echoes of Custer's loss at the Little Bighorn, appeared in the press, but matters soon calmed down. Bagone-giizhig avoided arrest; he had not even been at Sugar Point. In the end, no one was ever charged with a serious crime. The funeral of the slain soldiers was a major event in the Twin Cities. They had the dubious distinction of being the only troops killed in action buried at the Fort Snelling post cemetery.

⁓

Not long after the Sugar Point incident, the main body of the 3rd Regiment returned to Fort Snelling. It would be a brief stay, however. In January 1899, they were sent to the Philippines, not to return to Fort Snelling until 1921.

If any of the troops who had been part of the regiment in 1899 remained for the return to Fort Snelling in 1921, they might not have recognized the place, for the turn of the century saw a still greater expansion and rebuilding of the post. Among the critical new facilities was a modern hospital. The old building, erected in 1874, was "cheaply built," designed for a much smaller, four-company garrison, and, according to the 3rd Regiment's surgeon, W. C. Borden, "from its age and faults of construction beyond repair. It has no sewage system . . . no proper water system . . . the wards are badly lighted, badly heated, badly ventilated. . . . In fact the condition of the hospital is so wholly bad and unsanitary that it is almost impossible to point out the defects in detail."[33]

Funeral for soldiers killed at Sugar Point, 1898. The Department of Dakota's administration building is at the left. *MNHS*.

The army medical inspector provided additional details in a vivid description of the sanitary conditions. "Patients unable to leave the building use close stools. All patients who are able resort to an out house about 50 feet from the end of the west ward. . . . The excreta are removed daily by prisoners. . . . It is exceedingly disgusting to the senses and is primitive in the extreme."[34]

A larger hospital was proposed with every modern feature reflecting advances in medical practice, particularly as regarded sanitation. The new building, completed in 1898, brought health care at Fort Snelling to a standard as high as or higher than many military or civilian facilities.

Another ongoing challenge at the post was the location of its rifle range. The original range was constructed in 1880, but it had to be abandoned in 1891 because of encroaching civilian population and the fact that it was located on rented land. Weaponry had changed dramatically as well. The .30 caliber Krag-Jørgensen rifle, a more powerful, longer-ranged firearm than its predecessors, was adopted by the army in the early 1890s. Proper training for the effective use of the new rifle required a much larger practice range with

an artificial "butt" or berm behind the targets and other safety features intended to avoid what the medical examiner called "accidents that have occurred to citizens, from lack of proper precautionary construction."

Though the new range was located on the reserve, the army still had to work with Hennepin County to reroute several roads that ran through the fort. By 1903, a major expansion of the post was well underway. The water supply and sewage system were upgraded, and 857 acres of land were purchased along the north bank of the Minnesota River for the first expansion of the military reserve in over fifty years. With yet another shooting range located there, "there will be no need of marching down to Lake City for annual target practice, it is proposed to have an infantry range about two miles long." Most of the "new" property (it had all been part of the Fort Snelling reserve before 1858) was intended for training and maneuvers: "It will make a typical ground for sham battles, and will be extensive enough to give soldiers room to display their military skill." The army intended to house a full brigade of troops, including infantry, cavalry, and artillery, on the base. Amenities for the soldiers were also included in the new construction. A full gymnasium with a basement-level bowling alley and upstairs library and reading room, along with a new post exchange, would help morale and perhaps keep some of the men out of trouble in town.[35]

In 1904-5, an additional six sets of officers' quarters, a large riding hall for indoor cavalry drill, extension of the electric lighting system, and landscape improvements were all under construction. To go with all this expansion, Congressman Frederick Stevens of St. Paul, along with city business groups, began to push for a new, larger bridge linking St. Paul to the fort. In particular, they wished the new span to be large enough to support both road traffic and a streetcar line, tying the fort into the city's transportation system. When the bridge was completed in 1909, the roadway cut directly through the center of the fort's old parade ground; the streetcar line crossed the parade ground and turned to the northwest, past the round tower.[36]

While the "upper post" or "new post" seemed to grow exponentially to the north and west of the old fort, the army had not totally forgotten the original Fort Snelling. Though the old buildings were slowly disappearing—the last of the original stone walls was torn down in 1907—there were plans to reuse and upgrade some of the barracks, the commandant's house, and the round tower. By the turn of the century, as Minnesotans began to reflect on the state's past, the old fort with its associations to early Anglo-American "settlement" began to take on a nostalgic place in the minds of many citizens. This image was enhanced by newspaper coverage of Civil War regimental reunions, where surviving veterans of the Minnesota volunteer regiments recalled old times and former comrades in the shadow of the round tower.

When the army decided to renovate the old fort buildings with Spanish Colonial-style exteriors, the round tower was covered in a layer of stucco. The odd-looking results provoked a strong reaction from the local press. "Interesting Monument at Snelling is Mummified," complained the *St. Paul Globe*. "Famous Round Tower Typical of the Early Days the Object of Official Vandalism—shorn of its poetic vines and its old amber stone plastered with stucco, it loses all semblance of what it once was, and now resembles a big cake ready for the frosting." The *St. Paul Dispatch* cried,

A streetcar passes the round tower, about 1915. *MNHS*.

The reunion of the 1st Minnesota Volunteer Regiment, 1902. *MNHS.*

"They ensconced it with putty and made a venerable, beautiful fragment of the past into a cheese box." The *Minneapolis Journal* charged that the tower was "robbed of every suggestion of romance or historic interest by utilitarian military authorities." Even the ubiquitous Congressman Stevens made an official complaint.[37]

Army officials at first took the heat and stuck to their plan, but fortunately for all concerned, the old limestone walls did not support the stucco, which soon cracked and began to peel off, giving the army a face-saving excuse to remove it. This proved timely, as the tower was the focal point of a major commemoration of the centennial of Zebulon Pike's 1805 expedition. In September 1905, a "large number of Native Sons of Minnesota, Territorial Pioneers" along with the Colonial Dames, state dignitaries, and Professor William Watts Folwell, former president of the University

of Minnesota and chair of its political science department, dedicated a commemorative plaque at the tower. The army was represented by an impressive military contingent: four companies of the 4th Cavalry, Battery D of the 5th Artillery, and four companies of the 28th Infantry "stood at attention during the exercises."[38]

Clearly the old post had become something of a sacred site for white Minnesotans, a place to tell a story of successful expansion and rightful occupation. It was a symbol of what was remembered as a heroic and romantic past, celebrated without much comment on the suffering and loss of those they had displaced. "The history of Fort Snelling is one of peculiar interest," noted the *St. Paul Globe*, "especially to the people of Minnesota. . . . About the old stone tower and the buildings of the 'old post' are entwined memories that live in the hearts of the pioneers, and which like the

Civilians sketching the "romantic" round tower on an outing to Fort Snelling, about 1905. *MNHS.*

legends of the past, have a peculiar fascination for the younger generation." Minnesotans of European descent—the people making the decisions—were not ready to wrestle with the painful story of dispossession inflicted on the Dakota and the Ojibwe; after all, survivors of 1862 and their children still carried their own trauma.[39]

New construction and upgrades to older buildings continued through the first decade of the twentieth century. As new brick and stone structures appeared, many of the older wood-frame buildings added in the 1880s were pulled down. By 1910, the federal government had spent almost $1.5 million on Fort Snelling, a considerable investment, while the State of Minnesota, Ramsey County, and the Twin Cities Rapid Transit Company invested tens of thousands of dollars in infrastructure connecting the post to nearby communities. It seems odd, then, that in 1912, the US government seriously considered closing Fort Snelling.

For some time, the War Department had recognized the dramatically altered situation that now faced the army. For most of its existence,

the main task of the US Army had been to act as a frontier police force, securing and abetting American expansion along a nebulous, shifting border between European American "settlements" and Indigenous people who had "settled" their homelands for millennia. In doing so, the army helped to carry out a dubious and often corruptly administered Indian policy intended to bring "civilization" to Native Americans while taking over their lands. This mission had never required a large, permanent force, and the majority of Americans were not inclined to support one in any case.

By 1900, this frontier had long since ceased to exist. America's Indigenous people were living on reservations. On occasions, such as the War of 1812 or the Mexican-American War, the army had fought a foreign foe, and in the case of the Civil War a domestic rebellion. In each instance, it had experienced temporary expansion followed by immediate retrenchment. Troop numbers were reduced, unneeded posts abandoned. The organization of the old frontier army, created after the Civil War, no longer made much sense.

The Spanish-American War demonstrated that the old model of army expansion through the creation of volunteer regiments in each state was inefficient and ineffective. Bringing all of these relatively small and widely dispersed regiments together into an effective expeditionary force had proved almost unworkable. The regular army's organization and training also needed improvement. In response, the army tried to adopt a more modern division system made up of infantry and artillery brigades from different posts. The first attempt to employ this system came in 1910, when the army endeavored to mobilize a division to deal with problems on the US-Mexico border. It quickly turned into an organizational fiasco, and the entire effort was abandoned.

Starting in Theodore Roosevelt's administration, planning began on how the army should be organized and how it should interact with the states' National Guard units, created by law in 1903. This planning continued into William Howard Taft's presidency under Secretary of War Henry L. Stimson.

The "Stimson Plan," as it was known, organized the army into divisions and cavalry brigades "ready for immediate use as an expeditionary force or for other purposes." The continental United States was divided into four geographical regions, or "divisions": Eastern, Central, Western, and Southern. A regular army division was assigned to each geographic division. The National Guard was also grouped into divisions organized by geographical area.[40]

Minnesota fell into the Central Division. Headquartered in Chicago, it included the states of Ohio, Michigan, Indiana, Illinois, Wisconsin, Minnesota, North Dakota, South Dakota, Iowa, Missouri, Kansas, Nebraska, Wyoming (except that part included in Yellowstone Park), Colorado, and the post of Fort Missoula, Montana. This area had originally contained twenty-two army posts, and Stimson wanted to reduce the number to

nine. Though Fort Snelling was among the larger and more modern of the posts, there was still local concern. As in recent years, Representative Frederick Stevens took the lead in advocating for Fort Snelling. In a press interview, he summed up the situation: "The policy of the government in having only nine posts means that they must be used as training grounds for the army. In our effort to save Fort Snelling, it must be regarded not as a place for the stationing of troops, but as a place where troops can be instructed for service." He also felt sure the government would not walk away from the major investments it had recently made in the post's facilities.[41]

Stevens proved to be foresighted in his assessment. The recent investments in Fort Snelling guaranteed its survival, and over the remaining years of its service, Fort Snelling's role as a training and induction center would be increasingly important. Even in the decade prior to World War I, the fort hosted "sham battles," ran training maneuvers for Reserve Officers' Training Corps cadets from the University of Minnesota and other area colleges, and provided training facilities for the two National Guard artillery batteries located in the Twin Cities. The first real test of the Stimson Plan came in 1916. The political instability in Mexico had continued. Bands of guerrillas commanded by Pancho Villa raided towns in Arizona and Texas, killing American civilians and guardsmen. President Wilson decided to mobilize the army and National Guard to protect the border—and granted the army permission to follow Villa's men into Mexico if necessary. Regular army units from Fort Snelling, the 28th Infantry and 4th Artillery, went to the border. Minnesota Guard units, the 1st and 2nd Artillery Batteries, and the 1st and 3rd Infantry were mobilized at Fort Snelling and eventually went to Arizona as well. They never caught Pancho Villa, but the exercise showed the army that their new system still needed work.[42]

Fort Snelling, 1902. *Oil by McKelvey, MNHS; gift of Sibley House Association of the Minnesota Society of the Daughters of the American Revolution.*

11. World War

CHANGE WOULD NEED TO COME QUICKLY. THE US government had resisted involvement in the war raging in Europe since 1914. Woodrow Wilson had even campaigned for reelection in 1916 on the slogan, "He kept us out of war." But the scale of the conflict and strategies of the belligerents, particularly Germany's use of unrestricted submarine warfare, made neutrality impossible. The United States declared war on Germany in April 1917.

The Minnesota National Guard troops had barely scraped the mud of the Mexican border off their boots when they were called back to Fort Snelling for a more formidable task. Initially, the post resumed its role as a place for the mobilization of the recently released National Guard units back into federal service. It was apparent from the start of America's participation in the war that a much larger army than that represented by the regular army and National Guard would be needed. In May 1917, the US Congress approved a selective service act, requiring all men between the ages of twenty-one and thirty to register for the draft. Unlike the Civil War draft, there would be no substitutes, though some men, those with dependents or working in certain industries,

would be deferred. Sixteen sites located in the various geographic military divisions would be chosen as induction and training centers for the anticipated flood of volunteers and draftees. These cantonments would be enormous, containing hundreds of structures and miles of roads, sewers, water systems, and electrical utilities. For a time, Fort Snelling was among the locations considered in the Central Division, but the War Department settled on Des Moines, Iowa. The reason given was the lower price of land, though some state politicians believed the fact that Des Moines was in a "dry" county where alcohol was not sold, while St. Paul and Minneapolis were decidedly "wet," had tipped the balance to Iowa. Camp Dodge, as the Des Moines site was named, soon became the primary induction and training center for the Upper Midwest, including Minnesota.[1]

The vast expansion of the army required an equally large and rapid increase in the number of officers. There were not enough junior officers—captains, first and second lieutenants—then serving in the regular army or National Guard to fill the demands of the new formations. To meet this need, a number of posts were selected to host

Officer Training Camps, intended to provide a three-month crash course to young men desiring to be army officers, the so-called "ninety-day wonders." Fort Snelling proved to be an ideal location for such a camp, and during the war, more than 2,500 officers would train there. Thousands of young men, many of them college students or recent graduates, applied and took exams for entry into the program. "In general," noted a historian of Minnesota's participation in the war, "the qualifications demanded of the successful candidate at the close of his course were physical endurance and dignity of bearing, a proper measure of self-confidence, a knowledge of tactics, pedagogical skill, the ability to march long distances and to ride a horse, some knowledge of marksmanship, and a general familiarity with the principles of camp sanitation." Men of color were not admitted to the program or invited to apply. The army, like most of American society, remained segregated.[2]

The trainees were organized into provisional training regiments and given an intense introduction to military drill, both how to do it and how to command it, as well as extensive daily classroom work on all things tactical and administrative. These activities were topped off by daily calisthenics and biweekly cross-country marches in full combat gear. The officer candidates' days were typically eighteen hours long. In the last few weeks of training, there were fewer hours of classwork and many more in the field, digging trenches, learning how to fight in trenches, and experiencing the joys of exposure to rain, cold, and mud. It was an unprecedented method of instruction, beyond anything in the previous experience of the American army. Older men, those with physical limitations, and individuals lacking maturity often "washed out." As General William Sage, camp commandant, observed, there was no shame in this: "Seldom have men been subjected to such short, intensive and terrific training as in the effort to turn them into capable officers in 90 days."[3]

Between May and November 1917, two training camps were held at Fort Snelling. In total, more than 4,000 men were enrolled, and 2,500 of these graduated as commissioned officers. The great majority received commissions as lieutenants, but some more talented graduates were commissioned as captains, and a few really outstanding graduates earned the rank of major.

These young officers were training for a type of warfare unknown to the recruits who passed through Fort Snelling fifty years earlier. World War I saw the introduction of new weapon technologies, particularly poison gas, machine guns, and powerful, long-range artillery that utterly changed the nature of battle. Yet tactics were slow to shift. The huge, national armies mobilizing millions of soldiers continued to use mass assaults over open ground, creating staggering numbers of casualties. For example, the attacking British army at the Battle of the Somme in 1916 suffered almost 60,000 casualties in a single hour, slightly more than the total losses both the Union and Confederate armies endured fifty-three years earlier in three days of fighting at Gettysburg. Though the United States Expeditionary Force suffered far fewer casualties when compared to the armies of other nations involved in the conflict, American forces still took 320,000 casualties, including 53,402 men killed, in just two hundred days of combat. Another 63,114 died from disease, mostly influenza, during the war.[4]

Learning from the sad experience of the European powers, for whom the scale of killing and maiming came as a gruesome shock, the US government anticipated heavy losses from the beginning of its involvement in the conflict. In 1917, Congress authorized cooperation between the military and the American Red Cross to recruit and train nurses, while the American Medical Association assisted in recruiting qualified physicians and surgeons. By November 1918, the Army Medical Department had increased its hospital

At left, instruction in using a range finder to guide artillery, 1917 (*photo by Keystone View Company, Library of Congress*); below, other images of Reserve Officers' Training Corps activities, 1917 (*both MNHS*).

US Army General Hospital 29, about 1918. *MNHS.*

space in the United States to 120,000 beds and another 300,000 beds in Europe. They were needed. In all, the military would record more than four million hospital admissions during the course of the conflict.[5]

Fort Snelling became part of this expansion of medical facilities. The post hospital, which had been updated several times between 1870 and 1900, was larger and more modern than most. In September 1918, the Army Medical Department designated Fort Snelling as General Hospital 29. Though the Fort Snelling post hospital acted as the main hospital building, many other structures at the fort were taken over and converted to use for rehabilitation and educational facilities as well. Most of the fort essentially acted as General Hospital 29.

It served as the final stop in a long trail of treatment. The wounded were first brought to field dressing stations near the front lines, where they received initial attention; then to field hospitals further in the rear for more advanced treatments,

such as surgical procedures; and eventually to evacuation hospitals, where they received care that enabled them to be transported back to the United States. At a "debarkation hospital" in one of the major port cities, some of the less badly injured were discharged, but those needing further long-term treatment were sent by rail to a general hospital, like Fort Snelling, nearer their homes. Before the establishment of General Hospital 29, the base hospital at Fort Snelling had been dealing with the effects of the influenza epidemic that hit Minnesota particularly hard in the autumn of 1918. One newspaper report noted a thousand cases in Minneapolis, half of them soldiers. "This number included 510 army cases at the Fort Snelling hospital, and 70 known cases among civilian, and it was estimated by Dr. H. M. Guilford, city health commissioner, that fully 300 cases existing among civilians had not been reported to the city health department." A few days later, a paper reported the deaths of ten soldiers at the Fort Snelling hospital.[6]

The first trainload of wounded arrived on the last day of the war, November 10, 1918. As the band from Fort Snelling played "When Johnny Comes Marching Home Again," 134 wounded men disembarked at the Minneapolis depot. Though the fighting in France soon ended, the backlog of wounded continued to roll in for months. By December 1918, the lingering influenza cases, combined with the steady stream of wounded men arriving from France, pushed the hospital to its limits. The conversion of other buildings on the base began, and new facilities were constructed. Eventually, 1,200 beds for wounded and sick patients were created at the fort. General Hospital 29 was specifically designated as a "Reconstruction Hospital" for sick and wounded soldiers.

This meant that the Fort Snelling hospital, beyond treating wounds, would be involved in rehabilitation and recovery for individual soldiers.

The treatment received at General Hospital 29 represented the most advanced medical practice of its day. Just as weaponry had rapidly evolved after the Civil War, so had medicine. In the fifty years between 1865 and 1915, what was once an "art" had evolved into a science. Germ theory, anesthesia, and X-rays, combined with advanced and sophisticated surgical techniques, had completely changed medicine. Wounds that would have meant loss of a limb, permanent disability, or certain death to a Civil War soldier were now curable. The army, with the full support of the US government, was determined to do all it could

Wounded soldiers at General Hospital 29, February 1919. *MNHS.*

Nurse corps, General Hospital 29, about 1918. *MNHS*.

for their "boys." Though the treatments seemed miraculous, they could involve weeks or months of therapy. "It may require one slight operation or it may necessarily require several major operations. It may mean hundreds of treatments or massages, baths and electricity; it may call for months of painstaking effort on the patient's part, or it may mean finally the proper fitting of artificial hands or limbs. But no matter how great, if necessary it is done."[7]

Supporting this work was an organization remarkable for its efficiency and variety of treatments far beyond any facilities preceding it at Fort Snelling. There were, of course, surgeons, some of

whom had previously worked in field hospitals in England and France, where they had learned specialized techniques developed to deal with the terrible battlefield wounds common in the Great War. Most of these doctors were not regular army physicians but volunteers who had left often lucrative civilian practices to join the war effort. Lieutenant Colonel E. P. Quain, for example, was a native of Bismarck, North Dakota, and a graduate of the University of Minnesota Medical School who joined the Medical Reserve Corps in May 1917; he served at several military hospitals in France before his assignment to Fort Snelling in February 1919 as chief surgeon. Given the

hospital's designation as a reconstructive center, the most important physician was probably Lieutenant Colonel James Chapman Graves Jr. A graduate of Amherst and Harvard, Graves had also volunteered in May 1917 and spent twenty months working in military hospitals in both England and France, where he practiced his specialty in orthopedic surgery. He came to Fort Snelling in March 1919 as chief orthopedic surgeon.[8]

Nursing was critical to good care, and the general hospital had some sixty "trained nurses"—graduates of accredited nursing schools. While the surgeons might have the rank and prestige, these women provided the indispensable daily care vital to the recovery of sick or wounded soldiers. Women nurses were not new in either the

military or civilian context. The medical demands of the Civil War had led the army to organize a corps of some three thousand nurses, led by Dorothea Dix, who—according to the US Army Heritage and Education Center—required her recruits to be "between thirty-five and fifty years old, in good health, of high moral standards, not too attractive, and willing to dress plainly." Still, the army, being a socially conservative, if not backward, institution, had somewhat grudgingly accepted female nurses. The Spanish-American War seemed to change these attitudes, as women served courageously in disease-ridden hospitals in sometimes dangerous situations.[9]

This experience led to the creation of the US Army Nurse Corps in 1901. Though considered

Mess hall, Hospital 29, about 1918. *MNHS.*

part of the army, nurses held no rank, unlike their male counterparts in the Army Medical Corps. The women were required to be US citizens between the ages of twenty-five and thirty-five, unmarried, white, and graduates of a nursing school trained in theoretical and practical nursing. It was a relatively small organization. When America entered the Great War in April 1917, only 403 nurses were on active duty.[10]

These numbers would increase dramatically with the assistance of civilian hospitals and the American Red Cross. Within a year, nearly 12,000 regular army and reserve nurses were working in 198 stations around the world. By the war's end,

some 21,480 women would serve as army nurses at home and abroad. Many of the nurses at the general hospital had carried over from the Fort Snelling base hospital, where they had largely dealt with training injuries and the tide of influenza cases, both from Fort Snelling and other training bases in the region. Some of the nurses were victims of the epidemic as well. Their work, as described by one of the head nurses, Ida Merritt, was "not of the bright, illuminating kind"; often it consisted of tedious but necessary routines of taking temperatures, changing bandages, and encouraging patients, who were in pain or discomfort, while maintaining a cheerful, positive attitude. The best reward, according to Merritt, was witnessing the "gradual coming back to life of her patient."[11]

The hospital also contained a department specializing in the fitting and customizing of artificial limbs and other prosthetics. The Twin Cities had been a center for artificial limb manufacturing since the Civil War, and the general hospital may have benefited from this local expertise. While the staff treated "virtually every kind of wound known to war," they believed that the department could create a customized limb so good that "very few can tell the artificial leg from the natural when worn by an experienced lad."[12]

As part of its recovery function, the hospital also had an education department that taught academic subjects, mechanical skills, and crafts as part of patient rehabilitation. This group's aim extended far beyond basket weaving and included more formal education for those soldiers who were high school graduates, as well as technical skills such as typing or woodworking. Formed in December 1918, the staff of eighty viewed their work as partly therapeutic or "curative" and partly vocational, with a dose of entertainment in the form of movies and theatricals thrown in. All aimed at the overall goal of returning wounded soldiers to civilian life. This educational work was facilitated by a psychiatric team who evaluated each patient to help in his recovery. This aspect of treatment was without precedent in the past. Men who were less seriously wounded could enroll in classes teaching skills from auto repair to jewelry making. A staff of "Reconstruction Aids skilled in handicrafts" worked with men unable to get around. Projects often involved leatherworking, weaving, or macramé, giving some mental relief for men otherwise immobile with nothing to occupy their time. The projects provided physical therapy as well. One aide described how a soldier worked for hours a day on a loom he could operate with his good right hand but only stabilize with his damaged left hand; he gradually regained use of his injured hand.[13]

Supporting the work of the doctors and nurses was a detachment of the Army Medical Service. Three hundred to six hundred men, many of them draftees, performed a variety of duties. They cooked meals, provided clerical and administrative support (like keeping patient records and statistics), and even guarded the post. Before Fort Snelling was designated as a general hospital, the 36th Infantry regiment had been stationed there. With the expansion of the hospital in the fall and winter of 1918–19, that unit was transferred so some of their duties as guards for the hospital and the post were taken over by the medical service detail.[14]

For all the soldiers and workers at the hospital, the American Library Association created a hospital library staffed by volunteer librarians. This was but one of a number of community agencies that supported Fort Snelling's Hospital 29. Social agencies in particular, many of which had originally formed to help immigrants and promote social "uplift," turned their energies to supporting wounded soldiers. These included such disparate groups as the YMCA and the YWCA, the Knights of Columbus, and the Jewish Welfare Board. Before the war, they had seldom cooperated, but they freely did so in support of the hospital. The Jewish Welfare Board, for example, joined the American Red Cross to provide automobile rides for convalescing soldiers through city parks in Minneapolis and St. Paul, while the YMCA provided space for the Jewish Welfare Board–sponsored entertainments in the Y's gymnasium at the post.[15]

The work of these organizations reflected a wider public appreciation, at times bordering on reverence, toward wounded veterans.

Minneapolis Morning Tribune
November 17, 1918, p. 10

By Fred R. Coburn.

They played football at Northrop field yesterday and Minnesota's fighting-forces defeated their traditional rivals from Wisconsin in the wind and

Lounge area, American Red Cross Convalescent Building, Fort Snelling, 1918. The man at right selects a record to play on the wind-up phonograph. *MNHS.*

rain, 6 to 0. But it was not the hard fought game itself that made the afternoon one to be long treasured in memory. It wasn't the miserable weather, it wasn't the double band, and it wasn't any wild excitement over spectacular plays on the waterlogged gridiron. It's a Long, Thin Line of Khaki. With the game nearing the end of the first quarter 8,000 or more spectators saw a long, thin line of khaki hobbling through the gate, and come limping and stumping into the stands. But the football rooters knew them for what they were. Every man and woman in the enclosure stood erect, and every man bared his head and cheered and every woman clapped her hands. The game was forgotten for the moment. Nobody cared what the Maroon and Gold or the Cardinal players were doing out on the muddy field. They just watched the returned soldiers filing in and kept right on cheering. When the wounded lads had found seats the football populace sat back and turned a desultory interest to the game once more.

General hospitals were placed in various regions around the country so that recovering soldiers were nearer their homes, where families and friends might provide support in their recovery. Local newspaper reports from around Minnesota indicate that, to some extent, this idea seemed to work. The *Willmar Tribune* noted on January 16, 1919, "Olaus Anderson received a letter from his son, Private Paul A. Anderson, that he has arrived at Fort Snelling Hospital. He states that his injury includes the loss of his right arm. This news was quite a shock to his folks, who have not known what his injuries were. The father and brother Miner left for St. Paul today to visit with him." Another article in the same issue, headed "Gone to California," told another family's story: "Mr. and Mrs. Albert Ahlberg were down to Fort Snelling hospital last Wednesday to see Pvt. Reuben Larson. The mother, Mrs. Lars Olson, the brothers Edward and Henry and brother-in-law Albert Thorson, all of Colfax, were also down to see their soldier boy. His was a narrow escape from death. The spent machine gun bullet that injured him missed his grenade pocket by about an inch. Had it struck that he would have been blown to atoms. His wound has healed, but a case of pleurisy has developed, and he was taken to California Thursday with a number of other soldier invalids whom would be benefited by a warmer climate."[16]

Other stories were not nearly so upbeat. The *Little Falls Herald* reported, "Mr. and Mrs. Eugene Rasicot were at Minneapolis the first of the week for a visit with Ray Rasicot, who is at the Fort Snelling hospital for treatment. He was wounded in the head on October 3, and since that time has been confined to bed, the injury causing paralysis of the right side. He is slowly improving and will recover, it is hoped."[17]

With all of this effort, organization, and local support, it seems strange that in June 1919 the Army Medical Service ordered that General Hospital No. 29 close down effective August 1919. This was despite the fact that the month of June was one of the Fort Snelling hospital's busiest. By June 30, the *Minneapolis Morning Tribune* reported, "There are now about 1,000 patients at the hospital with 45 physicians, 110 nurses, 100 reconstruction aides and 600 enlisted men of the medical corps."[18]

Most of the patients were transferred to other facilities during the next few months, but some Minnesota veterans continued to receive care at the Fort Snelling post hospital. Beginning in 1921, this work was carried on by the US Veterans Bureau, which leased the Asbury Hospital in Minneapolis and the Aberdeen Hotel in St. Paul for the purpose. Following World War I and the demobilization of millions of soldiers, the role of veterans in American society became increasingly important, to a degree not seen since the Civil War era. The Veterans Bureau, later the Department of Veterans Affairs, would gradually become associated with Fort Snelling. In 1925, President Calvin Coolidge ordered the transfer of 160 acres from the Fort Snelling reservation to the Veterans Administration for the site of a new hospital. This eventually became the VA Medical Center complex.[19]

Though it operated for only eleven months, Fort Snelling's General Hospital 29 received 7,467 admissions. Its greatest number of patients at any one time was 1,761 in October 1918; these included more than 400 patients admitted from the fort, most of whom were likely influenza cases, but also 997 wounded men transferred from other hospitals. Buildings on the post had been renovated and new, temporary structures built to accommodate the Officers Training Center and later the General Hospital. These included a set of temporary, uninsulated barracks known on the post as "the Turkey Farm," which would continue in some form of use for the next twenty years.[20]

Relative quiet descended on the fort in the months after it ceased to serve as a hospital. Some reserve units occupied the post briefly before being discharged from active duty, and there was even some press speculation that Fort Snelling's days as an active post were numbered. But change was afoot once more within the US military, and new planning for the future role of the service would give the old fort an active part in the post-war army.[21]

Fort Snelling
APRIL 1935

PAUL KLAMMER

12. *The Country Club of the Army*

THE END OF THE GREAT WAR AND THE CLOSING of General Hospital 29 coincided with the one hundredth anniversary of Fort Snelling's establishment at Bdote. The Minnesota Territorial Pioneers Association, with the enthusiastic support of officers at the post, began planning for a commemoration of the event in August 1920. As with the observation in 1905, the remembrances were infused with nostalgia and did not attempt to reflect alternative perspectives. Events took place both at the fort and at the Minnesota State Fair, on its grounds between St. Paul and Minneapolis. Festivities at the fair began on September 4, 1921, with a pageant, including a presentation by Professor D. Lange, "writer of Indian stories for boys," who "related frontier tales" as the "Indian portion of the program." No actual Indigenous people seem to have been involved in this presentation.[1]

Another centennial ceremony, held at Fort Snelling's round tower, included a number of veterans and pioneer groups. Descendants of some of the important local figures active during the early years of the fort, carrying names like Steele, Botineau, Brown, and Larpenteur, were represented.

Curiously, there were no Snellings, Leavenworths, or Blisses among the military—and most certainly no Dakota representatives. Numerous speakers took part at both occasions, including the now very aged William Watts Folwell, Governors Joseph A. A. Burnquist and Samuel R. Van Sant, and a wide variety of clergy representing every major Christian denomination, each of which had its own day of commemoration at the state fair—an odd circumstance since missionaries played a minor role, at best, in the early history of Fort Snelling. Themes of their speeches seemed to focus more on Minnesota in 1920 than 1820: "Fort Snelling and the Present Generation," "The Value of Permanent Christian Organization," and "Minnesota's Position" appear to give scant attention to the actual history of the fort. Perhaps Dr. Folwell, who was then finishing his own comprehensive history of that era, touched on why and how Fort Snelling was built in his four-minute address to the crowd.[2]

Though some of the Twin Cities papers covered the centennial celebration, even devoting part of a Sunday edition to a historical overview of the establishment and building of Fort Snelling, other events probably dominated public attention. Even the issue of the *Minneapolis Tribune* with the most coverage of the history of Fort Snelling buried it behind "Freak Stuff at the State Fair"—auto polo, anyone? The same issue covered the ongoing presidential campaign, noting that all four candidates would be visiting the Minnesota State Fair. There may have been a certain nostalgic interest in the old fort, but it was not in the forefront of mainstream public attention.[3]

As its second century began, Fort Snelling found itself in a role far removed from its original mission. Westward expansion had become a distant memory, romanticized in literature and more recently in film, and the United States had become a major economic and industrial force, the recognized equal of the great European powers. By 1920, the United States had forcibly acquired colonies in the nearby Caribbean and the faraway Pacific, requiring a military presence to control and protect US interests. The Great War had compelled the army to mobilize the nation's economic and human resources in ways America's founders never imagined, creating an army of millions and transporting much of it overseas.

These changing circumstances drove the army to change, but like many institutions with a conservative and tradition-bound past, transformation came slowly and erratically. Nearly three million soldiers were demobilized in the first nine months following the end of the war. The question facing the US government and army leadership centered on how to maintain a relatively small army able to meet a national emergency but also capable of rapid expansion in the event of another large-scale, international conflict.[4]

In some form, this question had been examined within the army for decades. Beginning in the period following the Civil War, General William T. Sherman, commanding the army during Ulysses S. Grant's administration, sought a way for the army to maintain its technological and organizational competence, rather than regressing into its antebellum role as a kind of frontier police force. He asked General Emory Upton, a favored associate who had gained a reputation for innovation during the Civil War, to visit armies in Europe and Asia to examine their organization and tactics. When Upton visited Europe, the system created by the newly formed German Empire so impressed him that he went no further in his investigations. The German model, created first in Prussia, centered on a general staff, a collection of high-ranking officers acting as the "brains" of the army, controlling all aspects of military planning from logistics to fighting. Supporting it was a professional officer corps and highly trained troops. This relatively small force relied on a much larger reserve, made up of officers and men who had received training in the regular army.[5]

Upton produced an extensive report recommending that the United States adopt a similar system. Though he died in 1883, his report had great influence with the army's leadership, but critically it had little political support. The German system relied on conscription and direct army control over the reserves, two features unpopular with Congress and the American public. Some of the more powerful of the army's old guard, particularly Adjutant General Nelson Miles, resisted the idea of a general staff that would lessen their authority. So it was not until the end of the century and the debacles experienced during the Spanish-American War that Upton's ideas at last found political support. President William McKinley and his successor, Theodore Roosevelt, backed the reform proposals of Secretary of War Elihu Root. They succeeded in passing legislation creating a general staff, with a chief of staff acting as advisor to the secretary of war; a number of war colleges, providing advanced officer training and support for the general staff, were established as well.

Of particular relevance to Fort Snelling was

Root's attempt to modernize the National Guard and create a stronger connection between the guard and the regular army. The Dick Act, passed in 1903, required guard units to hold regular drill and target practice as well as an annual summer encampment for drill and training. The army was to assign officers to instruct guard units and hold regular inspections, and to offer further training opportunities to National Guard officers through the Army War Colleges (there were separate schools for infantry, artillery, and general staff). The Stimson plan, named for Taft's secretary of war, which designated Fort Snelling as a training base, was part of these efforts.[6]

But these reforms were slow to take hold. The states clung to the guard as their local militia, a source of police power and a field for political patronage. Federal interference here was unwelcome. Congress was also slow to provide funding for the extra guard training and equipment, while regular army officers had a poor opinion of their counterparts in the Army Reserve and National Guard. Early experiments with division-level maneuvers and the problems on the Mexican border in 1917 had illustrated these issues.

⁓

The United States' involvement in the Great War dramatically changed the situation, however. The value of quickly trained "citizen soldiers," often commanded by reserve officers or men trained in officer candidate schools like those run at Fort Snelling, had been proven in combat. In postwar planning, this experience informed the ideas of Colonel John McAuley Palmer and gained support in both Congress and the army. Palmer, a West Point graduate, had wide experience in the army, serving in overseas posts and as a member of the general staff prior to World War I. Here, he gained a reputation as a military planner, helping to create the Selective Draft Act of 1917. During the war, he served on the staff of General John J. Pershing,

commander of the American Expeditionary Forces in France. Palmer developed the idea of a "total army": a core professional army that would include the Army Reserve and National Guard. This concept strongly influenced the National Defense Act of 1920.

Though Congress had rejected the idea of universal military training, it accepted a reduced army of 280,000 with reliance on citizen soldiers who were to be prepared for war through the National Guard and an Organized Reserve. Officers, especially those preparing to join the Reserve, were to come from an expanded Reserve Officers' Training Corps (ROTC), a four-year program run by the army and navy to train college students—eligibility limited to white males at this time—to become Reserve officers. Officers also were to come from the Citizens' Military Training Camps, a kind of volunteer training program that originated in the Preparedness Movement prior to World War I. The Regular Army was to devote much of its energy to training both the National Guard and the Organized Reserve, but ideally these forces were eventually to be led by their own officers. Nonprofessional officers were to be offered opportunities for advanced training at army war colleges and granted tours of duty on the general staff. This plan would determine most of the military activities at Fort Snelling from 1921 to 1941.[7]

Another factor that would influence life at Fort Snelling between the world wars was the return of the 3rd Infantry in the autumn of 1921. The 3rd, now a "skeleton" regiment counting just three hundred soldiers, was ordered to march from Fort Sherman, in Chillicothe, Ohio, to Fort Snelling, a hike of eight hundred miles. "Red tape" was the official reason given for this unusual maneuver. According to press reports, the army did not have sufficient funds to transport the regiment by rail, though the cost of supplying the troops on the march exceeded what their train fare might have

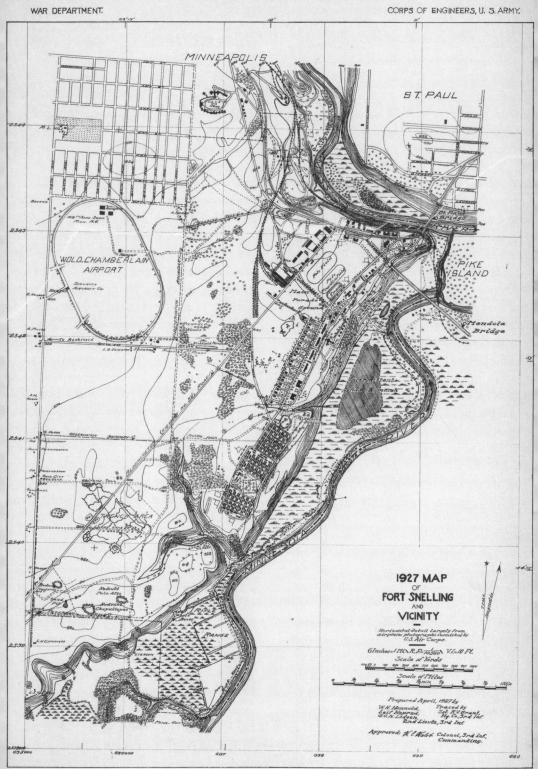

1927 MAP
OF
FORT SNELLING
AND
VICINITY

*Horizontal detail largely from
airplane photographs furnished by
U.S. Air Corps.*

6 Inches = 1 Mi. R.F. 1:10560   V.I. = 10 Ft.

Scale of Yards

Scale of Miles

Prepared April, 1927 by

W.H. Honnold,   Traced by
Leif Noprud,   Sgt. R.V. Grant,
G.O.N. Lodoen,   Hq. Co., 3rd Inf.
2nd Lieuts., 3rd Inf.

Approved: W.E. Welsh Colonel, 3rd Inf.,
Commanding.

*Reproduced from a 12" map prepared April, 1927
by W.H. Honnold, Leif Noprud, & G.O.N. Lodoen,
2nd Lieuts., 3rd Infantry.*
*Traced by Sgt. I.D. Dacquel, C.E., U.S.A.,
Office of the Engineer, 7th C.A., March 7, 1928.*

A 1927 map of Fort Snelling and its surroundings shows the tangle of transportation, polo fields, new airport, relocated firing ranges, entrenchment training grounds, and the press of the community on the old fort. *NARG 77, Records of the Office of the Chief of Engineers, War Department, Maps.*

been. But the march made for good publicity, and given that the US military had discovered the value of public relations during World War I, the 3rd's long walk may have been a convenient accident. Whatever the case, it made for a memorable return, and veterans of the trek celebrated the event for years, recalling mud, cold, and sore feet. Upon arrival at Fort Snelling in early December, they replaced the 49th Infantry and joined Battery F of the 14th Field Artillery and the 7th Tank Company as the garrison of Fort Snelling.[8]

The mission of Fort Snelling had evolved as well. The National Defense Act of 1920 created three branches of the army: the Regular Army, the National Guard, and the Organized Reserve. Regular Army posts were intended to be training sites for the National Guard and Organized Reserve with the Regular Army troops stationed at each post assisting with the training programs.

Fort Snelling was now within the 7th Corps area, a geographic designation roughly comparable to the departments created by the army in the late nineteenth century. The 7th Corps, headquartered in Omaha, included Minnesota, North and South Dakota, Iowa, Missouri, Nebraska, and Kansas. Most of the Guard and Reserve units training at Fort Snelling came from these states, particularly Iowa and North and South Dakota.

Another no-less-important role for the fort's garrison was to enhance the army's standing with the public. The 1920s were a challenging time for the armed forces and the army in particular. Though most of the public had initially supported America's entry into the Great War, its high cost in lives and treasure and the perceived failure of President Woodrow Wilson's postwar policies created feelings of disappointment and cynicism in many Americans. Some even saw it as a grotesque

waste of human lives for no purpose, except the enrichment of weapons manufacturers. So when Republican Warren Harding came into office in 1921, most Americans were happy to put the war behind them. Funding for the army was reduced in the name of economy, cutting back on both the training that had been proposed in the Defense Act and the size of the army itself. General prosperity and a large antiwar movement made it difficult for the army to recruit both enlisted men and officers. The army needed good public relations to ensure its viability and funding—and to pursue its efforts to become an army supported by citizen soldiers in the Organized Reserves and National Guard.

For its place in Minnesota and the Twin Cities in particular, the 3rd Infantry was an ideal ambassador for the army. The regiment had spent a decade at Fort Snelling between 1888 and 1898. Many of the men who had served in the 3rd during the Spanish-American War and the decade following had enlisted at Fort Snelling. Though few, if any, remained in the service, there were some veterans of the regiment in both Minneapolis and St. Paul. And if any unit in the army of that day could have been termed "elite" or "crack," it may well have been the 3rd Infantry. Proud of their history, "The Old Guard" (as they were and still are known) could put on an impressive military review. The troops were well drilled, their marching columns precise; as a nod to their history, the color guard wore the eighteenth-century uniforms of the early regular army. Their band was excellent. One frequent guest at post reviews who had substantial military experience in the British army wrote the post commander in 1931: "Last Monday's review was the best I have witnessed. . . . The Infantry marched well and showed solid and unbroken lines; the Artillery did better

The color guard of the 3rd Infantry Regiment, wearing their eighteenth-century uniforms, 1935. *MNHS.*

than ever, and that's saying a lot. . . . It happened that I took occasion to inspect the artillery and machine guns at close range . . . and was thrilled to note the cleanliness of their accoutrements, which compared most favorably with outfits with which I served so many years ago. . . . I always feel when troops make a good impression that a word of praise will not offend. Once a soldier always a soldier."[9]

This quality was put to good use in the 1920s and 1930s, both on the post and in the community. Though inspections and parades were a regular part of garrison life, the fort's were open to the public and frequently featured honored guests. Early in 1929, for example, Governor Theodore Christianson and 150 members of the Minnesota legislature visited the post. Battery F of the 18th Field Artillery welcomed them with a nineteen-gun

salute, after which all witnessed a "review of the command with winter equipment," followed by "a demonstration of a winter attack by a platoon of infantry supported by machine guns, skijoring, skating and hockey at the riding hall." In September of that year, the governor returned with Secretary of War James W. Good and fourteen others; all were honored at a review, made honorary members of the 3rd Infantry, "and presented with Swagger Sticks and certificates of membership."[10]

Other notable guests at the fort in 1929–30 included Minnesota's US Senator Henrik Shipstead and another prominent Minnesotan, former Secretary of State Frank B. Kellogg. Both were made honorary members of the 3rd Regiment. The army seems to have gone to particular lengths to honor Kellogg, who had negotiated several important peace agreements and in 1921 received the Nobel Peace Prize. Captain Glen R. Townsend, the regiment's information officer and the editor of the *Fort Snelling Bulletin* (the post newspaper), recognized him as "America's outstanding advocate of peace between nations. . . . The fact that he is, nevertheless, willing to accept honorary membership in the Army's oldest regiment should have a peculiar significance for those citizens of our country who confuse the cause of peace with the misguided (and often disloyal) pleas for total disarmament. There should be significance also that the United States Army through the Third Infantry takes this opportunity to pay tribute to Mr. Kellogg. . . . It is another

evidence of the fact that in all our national history the man in uniform has supported the efforts of those who honestly and wisely lead our nation in the paths of peace."[11]

Though these parades and demonstrations were useful for impressing special guests, the garrison also went to great lengths to perform for the general public. Beginning in 1928, the troops put on an annual "Military Show," featuring exhibits of aircraft, guns, tanks, and field kitchens and demonstrations of horsemanship, tank "acrobatics," and "silent drill" by the 3rd Regiment's drill team. The centerpiece for the whole three-day affair was a "mock battle" or reenactment, often

Mock battle staged at Fort Snelling, September 1928.
Lawrence Fuller Fort Snelling Research Materials, MNHS.

of a famous battle from the World War or a reenactment of the regiment's assault on Chapultepec during the Mexican War. These demonstrations, sometimes depicted in local newsreels, were full of troops firing blank bullets while charging through fake artillery bursts and other impressive but harmless explosions. The public loved it. Here was the excitement of battle without the blood, gore, and death. Some years, as many as 20,000 visitors from throughout the region attended.[12]

Perhaps the longest reach of the 3rd Infantry's showmanship came from its band. The participation of military bands from Fort Snelling had been a fixture in local events for years, but new technology made the 3rd Infantry band known over a wide region of the country. Radio was the exciting new medium of the 1920s, delivering entertainment directly into people's living rooms, often in the form of live performances. The Twin Cities were home to two "clear channel" stations, WCCO and KSTP, with powerful signals covering much of the Midwest. These stations were ever in need of quality live performances to fill up their programming schedules, and both stations turned to the 3rd Infantry band as a potential source. Starting in December 1928, KSTP began broadcasting a weekly Saturday evening concert of their music, which included band arrangements of classical pieces and waltzes. A month later, WCCO began broadcasting live concerts by the 3rd Infantry band on Friday evenings. Initially, those performances ended in April when the band's outdoor duties increased, but the public response to the broadcasts was so positive that the 3rd Regiment's commander, Colonel Walter C. Sweeney, arranged to have a radio broadcast studio set up in a little-used building on the post so the band could perform directly from Fort Snelling. The band's broadcasts earned praise from listeners as far away as Utah, but most came from locations "throughout Minnesota, Wisconsin and the Dakotas." These broadcasts continued into the mid-1930s.[13]

Like the special parades for VIPs, these shows and promotions were the brainchild of Colonel Sweeney, as was the *Fort Snelling Bulletin*, the post's weekly newspaper. Walter Sweeney was one of the army's rising officers. Originally a volunteer in the Spanish-American War, he earned a Reserve commission and later a Regular Army commission. He served on General Pershing's staff during the Mexican Border deployment in 1916–17, and during the Great War he served as chief of staff of the 28th Infantry Division. Sweeney also had a role in army censorship during the war and was instrumental in the creation of the army newspaper *Stars and Stripes*. He believed strongly in the value of military intelligence and propaganda, authoring a book on the subject in 1924; he was transferred from Fort Snelling in 1930.[14]

The regiment was also represented in parades and ceremonies outside the post by company- and even battalion-size participation. National holidays saw them in particular demand. In 1929, for example, the band, color guard, and three companies of the 3rd Infantry's 1st Battalion took part in the Minneapolis Memorial Day parade. Considering that this amounted to several hundred soldiers, they likely *were* the parade. At the same time, the 2nd Battalion was marching in the St. Paul parade. Earlier that day, the 3rd Battalion, with the band and color guard, had taken part in an elaborate ceremony at the Fort Snelling cemetery, and, not to be left out entirely, the artillery battery fired a "national salute" of twenty-one guns on the post-parade ground at noon. Even nonholiday events could draw participation from the garrison:

> Monday and Tuesday, April 29 and 30 [1929], a platoon of the Third Infantry with the Regimental Band will participate in ceremonies to be held in Minneapolis and St. Paul in connection with the shipment of a carload of material for the rebuilding of the frigate *Constitution*, normally known as

A unit of the 3rd Infantry marching perfectly in step in a parade along Sixth Street, St. Paul, 1929. *MNHS.*

"Old Ironsides." Monday afternoon also the garrison will hold a parade in honor of National Commander Grayson, Spanish American War Veterans. . . . Thursday, May 2, the Band, bugle and drum corps and a flag detail of non-commissioned officers will participate in the official opening of the baseball season in the Northwest at the American Association park in Minneapolis. Friday May 3, the garrison will be host to a party of Canadian Boy Scouts with their own band. . . . Friday May 10, Fort Snelling will entertain about 100 members of the Minnesota State Editorial association.[15]

These shows and parades demanded a good deal of time and effort from the soldiers. Long hours of drill and training combined with "spit and polish" were necessary. As one wrote on Colonel Sweeney's departure, "We have learned that we could efficiently shoot on the range, and at the same time carry on parades, put on Military Shows, paint and repair barracks, clean and generally renovate our buildings and surroundings." Yet life at Fort Snelling was not all work. Since the late nineteenth century, the army had shown increasing concern for the welfare of soldiers, first by increasing regulation of post sutlers in the 1870s, then replacing them altogether with post exchanges, the proceeds from which funded recreational amenities. In 1903, Congress made funding such facilities and related activities part of the army's regular appropriation. These resources allowed the building of facilities like a library, gymnasium, and post exchange (actually the old military prison converted for the purpose), all added before the World War. When the post expanded during that conflict, many amusements quite unknown to most of their predecessors were made available to soldiers through social agencies and later the army itself. In 1918, the Army Morale Division was created, followed in 1920 by the Army Motion Picture Service and in 1923 by the Army Library Service. All of these amenities aimed at improving the lives of enlisted men and their families to heighten morale and encourage reenlistment.[16]

A full range of sports was available for participation and entertainment. Fort Snelling had a football team that regularly competed not only with other military posts but with local high school and college squads. Its baseball team played against the many town teams around the state, and in winter, that most versatile of buildings, the Riding Hall—a structure too large to heat—was partially flooded and used for ice skating, curling, and hockey. Basketball was also popular, with intramural competitions between the various companies. Boxing was particularly well liked. The sport was widely followed in the United States during the 1920s and 1930s, and the fort had an active league competing among the commands within the post and within the army as well. The Fort Snelling boxers also regularly competed in amateur bouts in the Twin Cities in fight "cards" or programs that attracted large crowds and garnered wide coverage in the local press. For those of more refined taste, perhaps, the fort also provided a golf course, completed in April 1932. Soldiers could play up to thirty "rounds" for one dollar. The course offered many of the features of a country club course, with a golf director determining handicaps and a fully equipped golf shop where clubs and golf balls could be purchased. Indeed, when the officers' club reorganized itself as the Fort Snelling Country Club in 1934, the post had truly become "the country club of the army."[17]

The post movie theater, built in 1917, offered a full range of feature films. In early 1930, the post theater received a major upgrade when nearly $5,000 worth of equipment (the equivalent of roughly $77,000 now) was purchased and installed to improve the playing of "talkies," movies with sound. "A new and specially prepared screen made for talking pictures arrived this week, [which] will reflect the pictures better and is several feet bigger

The lounge of the Fort Snelling Officers' Club, reorganized into the Fort Snelling Country Club, August 1934. The wooden beams were salvaged from the old Minneapolis Armory. *Lawrence Fuller Fort Snelling Research Materials, MNHS.*

"To the Army Mule"

In February 1934, the Fort Snelling *Bulletin* published an ode celebrating pack animals so patient they might give a cat a ride.

Oh mule, You are a wondrous thing, No horns
    to honk, no bells to ring.
No spark to miss, no gears to strip; you start
    yourself, no clutch to slip;
No school to go to four times a week to try and
    learn from where you squeak. . . .
Your frame is good for many a mile; your body
    never changes style.
Your wants are few and easily met—you've
    something on the Motor yet.[i]

as the 'sound box' of the talking pictures is behind the screen." A new and completely up-to-date movie theater was built in 1933. As with other cinemas of the day, air-conditioning contributed to its popularity.[18]

But the post theater was not the only movie house available. The whole of both Minneapolis and St. Paul were accessible to members of the garrison and their families. Soldiers could now travel more easily than in the past. Streetcar lines connected the post with both nearby downtowns, and many in the garrison owned automobiles as well. The Fort Snelling garrison enjoyed a closer physical connection with the cities than it ever had in the past.[19]

The social connections were stronger as well, and perhaps no links between fort society and Twin Cities society were more in view than those involving horses and horsemanship. Horses and mules continued to be important modes of army transport in the years following World War I. The army viewed them as more reliable than motor vehicles in battlefield conditions and away from roads.

But for officers in the peacetime army of the 1920s and early 1930s, the horse remained part of their profession, a sign of status and an enjoyable avocation, as well as an integral part of army culture. They and their families took part in riding schools and riding shows at the fort. The shows were big public events, with hundreds of "St. Paul and Minneapolis society leaders and horse lovers" attending. The annual Military and Civilian Horse Show featured equestrian competition among civilian and military riders, including women, from the Twin Cities and Fort Snelling. The events were popular with the riding public, often attracting hundreds of entrants. Later, a Spring Junior Horse Show was added, with competitors from the various children's riding classes; some of the events were open to youngsters from the Twin Cities and vicinity.[20]

The real equestrian star at Fort Snelling was an unlikely looking chestnut stallion affectionately known as Whiskey. He was a wild horse from Montana, acquired by an army remount station and sent to Fort Snelling in 1921. Whiskey was supposedly so troublesome and ungainly that he would have been destroyed had not a young lieutenant named William Hazelrigg taken an interest in him. Hazelrigg had a gift for training horses and soon developed a bond with the horse. First trained as a polo pony, Whiskey proved a natural at the sport and became a star of the Fort Snelling team. Realizing Whiskey's intelligence, Hazelrigg gradually taught him an array of tricks. Whiskey eventually became adept at picking up handkerchiefs, tipping the lieutenant's hat, bowing, or sitting on his haunches like a dog. At horse shows and equestrian competitions, Whiskey soon became a crowd favorite. Besides the usual show jumping, at which he excelled, he performed an array of stunts, including jumping over a mule standing between two fences, a dinner table with diners in their seats, or a flaming fence.

Lieutenant Hazelrigg was transferred to a post in the Philippines in 1926, and Whiskey was left behind. But the horse's career continued with a variety of riders, among them the wives of officers at the fort. He was so popular at Fort Snelling that he was allowed to roam the post at will during the day, grazing where he liked, and he frequently left his stall at night (he figured out how to open it) to wander, startling sentries. He continued to perform at shows until he was officially retired in 1934. Even then, he occasionally came out of retirement; as late as 1941 he was doing his bit to entertain trainees at the Fort Snelling Reception Station. Whiskey died on December 30, 1943, and was buried with full military honors on January 1, 1944. He was the last government-owned horse at Fort Snelling.[ii]

Whiskey, the show horse, jumped over tables, other horses, and, here, trusting servicemen, about 1930. *MNHS*.

There was also a Fort Snelling Hunt—not one involving an actual fox, but a drag hunt, wherein a heavily scented substance, like a bag of anise seeds, is dragged behind a horse, creating a scent trail for the hounds to follow. This was generally a set course through the undeveloped areas of the Fort Snelling reservation, with sufficient jumps to challenge riders and their mounts. Women were among the riders, and a number of hunts were regularly opened to riders from around the Twin Cities area.[21]

The star equestrian event, however, was polo. An ancient game adopted by the British army in India in the nineteenth century, the sport became popular among the upper classes in Britain and later in the United States. Polo demands riding skill as well as highly trained horses. For some American army officers the sport became a way to preserve horse culture and horsemanship within the army. Fort Snelling had several teams that competed against other army posts and local

Fort Snelling hunt, 1932. The hounds chased a scent, rather than a fox. *Lawrence Fuller Fort Snelling Research Materials, MNHS.*

Polo players, about 1931. *MNHS.*

civilian teams. Meets at Fort Snelling drew large civilian crowds and acquired a certain cachet within Twin Cities and area society. As one officer described it, polo "gives the zest of competition between teams and is not only preserved from the absurdity of undue importance, but is justified by a kind of romance which animates it."[22]

This comment may explain why polo was covered in both the sports sections and society pages of local newspapers: "The rapidly increasing number of polo enthusiasts in the Twin Cities will be pleased to hear that plans are being made at Fort Snelling to open the polo season. . . . Season parking places from which to view the games are already on sale by the Ft. Snelling officers 'Club.'" Most of the equestrian events seemed to appear on the society pages; the same pages also offered regular coverage of other happenings at Fort Snelling, as well as the comings and goings of officers and their families.[23]

These amusements and diversions at Fort Snelling came in the midst of a world and national crisis, the Great Depression. Between 1929 and the end of 1932, the army had little to do with the government's handling of the situation, with the unhappy exception of the Bonus Army marchers. In the summer of 1932 several thousand unemployed World War I veterans marched on Washington demanding that Congress grant them a bonus payment for their service, as had been done for veterans of past conflicts. The proposed legislation did not pass, but the Bonus Army, as it was called, remained in its encampment near the Capitol. When city police tried to remove them, a riot ensued and a number of people were killed. President Herbert Hoover ordered the army to clear out the camp. About six hundred cavalry and infantry under the direct command of Chief of Staff General Douglas MacArthur drove out the marchers and destroyed their camp. These events were widely covered by newspapers and movie newsreels. The army was following the orders of the president, but most of the public thought they had acted in an overly violent and heavy-handed manner.[24]

Following the election of Franklin D. Roosevelt in 1932, the US military was called upon to directly assist in relief efforts, most conspicuously with the organizing and establishment of the Civilian Conservation Corps (CCC) and its camps. In 1933, Congress passed the Emergency Conservation Work Act, establishing the CCC. Its goal was to provide the nation's masses of unemployed young men work on much-needed conservation and reforestation projects. Though army leaders complained that the task would interfere with training of soldiers, President Roosevelt ordered the army to mobilize the CCC and run its camps. Across the country, army officers and soldiers seem to have carried out the assignment with goodwill, enthusiasm, and efficiency. It's not hard to imagine that

the troops identified with the young CCC recruits and were happy to help them. Some 310,000 men were organized into 1,315 camps within the first seven weeks of the program. This was the fastest, most efficient mobilization in the history of the United States Army.[25]

The northern forests of Minnesota and Wisconsin made the area an important region for Civilian Conservation Corps activity. Fort Snelling had the formidable task of recruiting, organizing, and sending north fifteen CCC companies by June 1, 1933. Since each company or camp included around two hundred men, that meant processing three thousand recruits in two months. The job was somewhat complicated by the fact that the army was strictly forbidden to "militarize" the recruits. They seem to have avoided this by dispensing with formalities like rank, drill, formations, and military etiquette, while insisting on cooperation and good behavior from the recruits. Since most recruits were willing and even enthusiastic, the system worked well.[26]

The CCC was intended to be integrated, but African Americans enrolling in the program found that it followed the racist, segregated army system. Black workers, who made up only ten percent of recruits, usually found themselves in segregated camps such as those at Fort Snelling State Park and Temperance River. The few African American recruits sent to white camps lived in segregated barracks and had to eat apart from the rest of the camp. Beginning in 1933, CCC camps for Native Americans were set up in northern Minnesota, many of them near existing Ojibwe reservations. Unlike the other camps, these were administered not by the army but by the Bureau of Indian Affairs. Many of their projects focused on preserving Dakota and Ojibwe culture.[27]

A CCC camp was set up at Fort Snelling in April 1933 in the Citizens Military Training Corps cantonment. Eight officers and thirty-five enlisted men were assigned to the camp, which

A group of Civilian Conservation Corps members leaving
Fort Snelling for the Superior National Forest, 1933. *MNHS.*

acted as a kind of induction center. CCC recruits were treated somewhat like draftees. On arrival, they were registered, bathed, given a physical exam, and issued clothing. Recruits later received a certain amount of basic training: calisthenics to "build them up" and instruction in how to properly use the tools they would need in forestry work. They also learned basic hygiene, first aid, "and ready cooperation with those placed in charge of them." Recognizing how hard the lives of some of these young men had been, the army allowed that "if a candidate appeared undernourished he was fed before being processed." The military personnel also identified recruits who might make good supervisors "in order that they will become self sustaining." While the military organized all of this, it was the CCC men themselves who made the system work: "The ready willingness with which these men have cooperated with our officers and enlisted men has made the work a pleasure to our personnel. They appear grateful for the careful attention given them."[28]

As each camp was organized and sent off to its intended destination, detachments of officers and enlisted men from Fort Snelling went with them. The soldiers remained at the camps until they were well established. For the most part, the military men seemed to enjoy the change of routine and the experience of being "up north." Captain H. D. Dinsmore reported to the *Fort Snelling Bulletin* from CCC Company 703 at Caribou Lake that the "old fishermen from the post" were catching their limit of fish, and one of the sergeants "is considering buying some land up here . . . because he likes the country so well." Given that each camp was supervised by an army captain with two or more noncommissioned officers (NCOs) and perhaps five or six privates, the sixty-two camps operating at the end of 1934 required an extensive commitment from Fort Snelling. At the post itself, the CCC activity had converted Fort Snelling into "a small metropolis." Between April 1933 and April 1934, 19,500 enrollees were processed through the fort, while 11,773 had been discharged at the end of their term. At the same time, the CCC quartermaster, also operating through Fort Snelling, was supplying, each day, "18,000 pounds of potatoes, 20,000 pounds of beef, 6,000 quarts

of milk, 14,000 loaves of bread and 60 cases of eggs and other items in similar proportions" to feed the men in the camps. This period represented the height of the military's involvement. Over the next three years, the CCC did indeed become more self-sustaining, but Fort Snelling continued to process enrollees for several more years.[29]

Another New Deal agency that had an impact on Fort Snelling was the Works Progress Administration (WPA). In 1935, a combined grant from the WPA and army provided over $1.6 million for construction, modernization, and improvements at the post. Workers drained two swamps, constructed a new rifle range, refurbished and modernized barracks, and regraded and improved the parade ground. They laid new water mains and took on other infrastructure projects as well. Many of the flimsy wooden buildings built in 1917—the "Turkey Farm" barracks—were torn down to be replaced by new facilities.[30]

This work overlapped somewhat with the creation of the Fort Snelling National Cemetery. Though it was approved as a WPA project in 1936, the location of a cemetery on that part of the military reserve began a decade earlier, when the former post cemetery was moved in 1926 to make way for a new chapel. The new cemetery included the largely unmarked graves of soldiers who had died at the post over the previous century.

Interred there as well was the daughter of Abigail and Josiah Snelling, Elizabeth Snelling, who died in 1821. The national cemetery was dedicated on July 14, 1939, though the first burial, that of Medal of Honor recipient Captain George J. Mallon, took place on July 5.[31]

Since the 1920s, the units at Fort Snelling had made their regular summer "hikes" to training areas for maneuvers: the 3rd Infantry with the Tank Company to Camp Ripley and Battery F to Camp McCoy, Wisconsin. Though the training was as thorough as officers could make it, there was a certain routine, particularly as the troops were still working with weapons from the Great War. But in 1935, more ROTC and Reserve Officers were assigned to the post for training, and in 1936 the Camp Ripley maneuvers included the entire garrison and the brigade of which they were a part. Reserve Officers also trained at Fort Snelling by taking the place of regular officers during drills and exercises. Later in the decade, events overseas would begin to change these routines dramatically.[32]

Another regular part of the training regime at Fort Snelling was the Citizens' Military Training Camps (CMTC). In this voluntary arrangement civilians signed up, at their own expense, to undergo military training during the summer. Participants, who were mostly in their late teens and early twenties, learned military discipline,

OPPOSITE, LEFT Old Fort Snelling cemetery, 1905. *MNHS.*

OPPOSITE, RIGHT Burial plot of unknown soldiers, Fort Snelling National Cemetery, 1938. *MNHS.*

LEFT Civilian Military Training Corps troops march in review, July 1937. *Lawrence Fuller Fort Snelling Research Materials, MNHS.*

organization, and the basics of drill and tactics. They even learned to handle a rifle on the shooting range. In some ways, it was like the army's basic training—and in others, rather like a Boy Scout camp, for there was plenty of recreation and competition. Some of the initial training came from regular army officers and NCOs from the 3rd Infantry, but beginning in 1929, Reserve Officers began to run the training program, thus giving them useful experience.[33]

This program grew out of the "Preparedness" movement prior to the World War and was a favorite of General Leonard Wood and General John McAuley Palmer. Thus the CMTC remained a part of the army's training regime, though its participants were under no obligation to actually enlist. Most summers at Fort Snelling saw around a thousand participants. This degree of participation suggests that for some of the population, the military and national defense remained popular. Like the CCC, the CMTC proved a useful introduction to army life for young men who would later serve in World War II, and it even provided a source of Reserve Officers.[34]

The later 1930s saw many changes at Fort Snelling beyond the physical improvements to buildings and grounds. The rise of fascism in Europe and the rearming of Germany under Adolf Hitler created growing apprehension in the United States. Although Congress was leery of foreign commitments, it gradually increased funding for the army, enlarging the size of the standing army and authorizing more training for the National Guard and Reserve. In April 1939, General George C. Marshall became army chief of staff. He began a much-needed reorganization of the army and advocated for the mechanization of the entire force. The days of the horse and mule were over. At Fort Snelling, the era ended tragically. On April 9, 1939, a fire broke out in the stables of Battery F. The blaze spread quickly, killing 129 of the battery's 131 horses. They were never replaced; instead, the battery itself was disbanded and reformed as Battery A of the motorized 79th Field Artillery. That same year, the 7th Tank Company was reorganized as an infantry unit and sent elsewhere.[35]

When war broke out in Europe in September 1939, President Roosevelt declared a "limited" national emergency. The size of the army was increased to 227,000. A number of Organized Reserve and National Guard troops were called to active duty, and training of the Regular Army intensified. Some of the units in the Fort Snelling garrison were called to other posts, and Fort Snelling would find itself in a new role in a new World War.

Fort Snelling round tower, 1949. *Watercolor by Jo Lutz Rollins, MNHS; gift of the artist.*

13. World War II

MANY AMERICANS LIVING IN 1939 AND 1940 viewed the war in Europe like some distant storm: terrible and threatening, its path uncertain. They hoped it would pass them by. But as the news grew more menacing, preparation seemed increasingly wise.

President Franklin Roosevelt certainly realized the danger, but his decision to run for an unprecedented third term caused him to tread carefully. Isolationist sentiment in the country was still strong. Stunning German victories in Europe, however, especially the Blitzkrieg attack that shattered France in the summer of 1940, quickly altered public opinion and that of Congress as well. On September 16, 1940, Congress passed and President Roosevelt signed the first peacetime draft law in American history, the Selective Training and Service Act of 1940. Men between the ages of twenty-one and thirty-six were required to register for selective service, and the following month draftees, initially called up for one year, were streaming into reception stations.[1]

Fort Snelling was already gearing up to accept new enlistees. In August 1940, the *Minneapolis Tribune* reported on new construction at the fort of buildings to receive recruits and, since the draft bill was making its way through Congress at the time, men who might be drafted. The post became Reception Station No. 16, one of thirty-three in the country designated to receive and assign new recruits. By early November, work on new barracks was being rushed to completion. Located between the old Upper Post and the ROTC cantonment, the new buildings took over the area where the temporary World War I "Turkey Farm" barracks, demolished by the WPA, had stood. The WPA's extensive upgrades to Fort Snelling showed that the army had been planning ahead.[2]

Training intensified in 1940 and 1941 for the troops already stationed at Fort Snelling. The 3rd Infantry, as part of the 6th Division, took part in large-scale exercises in far-off locations like Nevada and Louisiana, returning for a time to Fort Snelling, where they helped train new draftees.[3]

After the excitement of the renewed draft at the end of 1940, 1941 must have been a year of odd suspense. As fighting raged overseas in North Africa, the Balkans, and China, America caught its breath. There was certainly support for national defense and for helping Britain in its fight against

In October 1940, as the first seventeen buildings of Reception Station No. 16 neared completion, the War Department announced that ten more barracks, three mess halls, and additional buildings would be added. *Lawrence Fuller Fort Snelling Research Materials, MNHS.*

Reception center barracks, 1941. *Lawrence Fuller Fort Snelling Research Materials, MNHS.*

Bugle calls were amplified through a megaphone, 1941.
MNHS.

Nazi Germany, but there was little enthusiasm for active intervention overseas. And while much had changed at Fort Snelling in two years, some of the old country club culture lingered. A large military police battalion, the 701st, arrived on the post, and the reserve 35th Combat Engineers regiment was activated at Fort Snelling in June. The post must have been bursting at the seams, but polo went on. "Fort Snelling Polo Brings Out Society" was the headline on the *Minneapolis Tribune*'s Sunday society page on July 13, 1941: "Society from the Twin Cities and neighboring states will be out for the games. . . . Wives and members of the visiting teams will be entertained at many parties being planned."[4]

Some draftees originally called up for a year were anxious to get back to their civilian lives. At a few posts, but apparently not at Fort Snelling, disaffected men were writing the word OHIO, meaning "Over the Hill in October," on barracks walls, expressing their determination not to reenlist after their original one year of service expired. At the urging of military leaders, Congress voted to extend the enlistments, but the legislation passed

by the slenderest of margins: a single vote. Again, events overseas altered the situation in America. In September, the German army invaded Russia. And on December 7, the Japanese navy attacked the US Pacific Fleet at Pearl Harbor in Hawaii. The following day, the United States declared war on Japan, and two days later, Germany declared war on the United States. In the words of British Prime Minister Winston Churchill, "The United States was in the war, up to the neck and in to the death."[5]

According to the *Fort Snelling Bulletin*, news of the Japanese attack on Pearl Harbor was met with cool resolve: "Orders issued by Colonel Harry J. Keeley, Fort Snelling's Commanding Officer, were summarily executed. There was no confusion no questioning. Action was and will continue to be the answer of the Post's officers and men to Army orders. . . . Fort Snelling's atmosphere remained calm, but alert." The *Minneapolis Tribune* had a different version. "Fort Snelling Tense," the headline read on December 8. "Post officers said men at the fort were ready for any action demanded of them and that what troops are there could be ready to move in very few hours." Yet no leaves had been cancelled, and no troops on furlough were recalled to the fort. The relatively low-key reaction probably stemmed from discipline and the reality that Fort Snelling was already performing an important wartime task, receiving and assigning draftees and recruits.[6]

Most recruits came by train or bus, though Twin Cities men might arrive via streetcar. Sometimes they traveled in a group, following a big send-off from their hometown. The reception at Fort Snelling was likely less uplifting, as new arrivals were marched off to spartan barracks under the supervision of a noncommissioned officer. On their first day of this new life, the recruits or draftees were given a thorough physical exam and then a battery of tests to determine their assignment to one of the three branches of the army: infantry, artillery, or air corps. Then the men

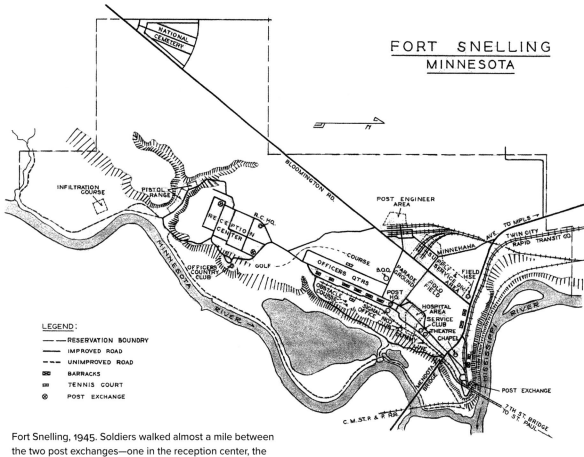

Fort Snelling, 1945. Soldiers walked almost a mile between the two post exchanges—one in the reception center, the other near the old fort. *Fort Snelling Map Collection, MNHS.*

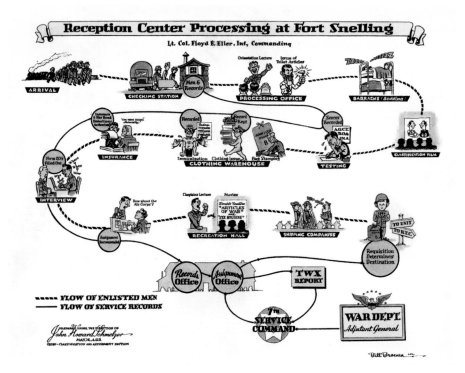

Recruits and draftees received a pamphlet with a cheerful graphic showing the process, about 1943. *MNHS.*

were issued uniforms and received inoculations for tetanus, smallpox, and typhoid. An army classification officer interviewed each enlisted man, asking about educational background, civilian job training, or any other experience that might influence his placement. The army needed mechanics, cooks, and electricians, of course, but also balloon riggers, tuba players, third rail repairmen, and seemingly every other specialty. In the words of the reception station commander, Lieutenant Colonel Floyd Eller, the intent of the screening was "to provide each army organization with men that it can use, train and keep."[7]

Once test results and interviews were completed, the new GIs, organized into temporary companies, began to experience the routine of army life as they waited for assignment: taking shifts on KP (kitchen police), cleaning barracks and latrines, and learning the basics of marching and drill. A recruit might remain at Fort Snelling for a few days or a few weeks before receiving orders to report for additional training at another post or to join a unit as a replacement. This system was affected by the reality of where the army actually needed men. Once American troops were in extensive combat in Europe and experiencing many casualties, particularly in 1944, the majority of draftees went to the infantry.

———

During the war, the Fort Snelling Reception Station processed approximately 250,000 recruits and draftees. At the opening of the Recruit Reception Center in September 1940, the center was run by a few Regular Army and Organized Reserve officers and 47 enlisted men from the 3rd Infantry. But with the start of the draft, more personnel were added: there were 83 officers, 718 enlisted men, and 110 civilians during the height of activity in 1942. In March of that year, 13,000 men passed through the station as they were inducted into the army.[8]

Meanwhile, the 1st and 3rd Battalions of the 3rd Infantry were deployed to Newfoundland in 1941, where they remained until 1943. The 2nd Battalion continued at Fort Snelling, acting as a training cadre. In September 1943, the entire regiment was sent to Camp Butner, North Carolina, and eventually shipped out to France in 1945, where they joined the 106th Division. With the departure of the 3rd Infantry, the last regular garrison troops left Fort Snelling.[9]

Even before their departure, the character of the post had changed drastically. The placid, orderly pace of life at the interwar post had been overwhelmed by the arrival of thousands of transient enlistees. Most of these new soldiers spent only a few days or weeks at the post before orders sent them off to other assignments, but they were able to enjoy many of the same amenities that the former "country club of the army" had offered.

Some sexual activity among soldiers has likely occurred at Fort Snelling from its earliest years to the present, but it is difficult to document. Until the late nineteenth century, the concept of homosexuality as a behavior among men or women was not defined—and people keep private lives private. Only after the 1960s, when civil rights movements included growing activism among lesbian, gay, bisexual, and transgender people, did a story emerge. Gay rights activist Chuck Rowland, who served at Fort Snelling during World War II, recalled in 1984 that Fort Snelling's induction center was quietly known as a gay-friendly workplace: "With the exception of the Master Sergeant, who ran this whole vast operation, every single person—corporals, sergeants, staff sergeant, tech sergeant, all the department heads were gay and were all guys in the single gay bar in Minneapolis [the Onyx Bar], . . . and they called [Fort Snelling] the 'Seduction Station.'" When recruits were forming a line outside the office windows, "one of the guys would say, 'Hey, hey, come look. Look, look at those beautiful young things going over to Europe to be killed. Terrible, terrible thing.'"[i]

The post exchange, library, and movie theater—and above all, dances—seemed favorite distractions, with young women from the local community taking part. On January 9, 1942, the *Fort Snelling Bulletin* reported that "500 girls . . . who are members of St. Paul Winter Carnival Units, will serve as partners for soldiers of the Post for the largest and best dance ever to be staged at Fort Snelling." Streetcar lines gave access to the downtown centers of the Twin Cities, where movies, clubs, and bars were ready to entertain. Both cities also had large United Service Organizations (USO) facilities that provided entertainment and support as well.[10]

The post continued to publish the *Fort Snelling Bulletin*, though the content changed noticeably. There was a greater emphasis on entertainment for enlisted men, what was going on and where to find it, and humor as well. Wartime censorship probably constrained any news content; there was no mention of the 3rd Infantry's departure, for example. But the conversion of the riding hall

United Service Organizations volunteers with servicemen at Fort Snelling, 1942. *MNHS.*

into a dance venue was a front-page item, as was the opening of the Service Men's Center in Minneapolis. One *Bulletin* column listed the Riding Hall dances "featuring Sev Olson and his popular orchestra," movies at the post theater (including the 1941 Oscar winner *How Green Was My Valley*), the "Sapia Swing Review" (a USO production "gathering plaudits where ever it has shown"), and finally, free tickets to the Shrine Circus. "We don't know how your entertainment tastes ran in civilian life," the writer continued, "but we do know that for a civilian to enjoy the entertainment we have at Fort Snelling this week would take a neat chunk of coin at civilian prices."[11]

Military humor and cartoons had always been a feature of the *Bulletin*, but they reached a peak during the early years of World War II. Talented cartoonists usually poked fun at the army and often featured attractive women. Later issues, beginning in 1945, became much more professional in appearance, with the production quality of a true newspaper, including photographs. They often featured stories on individuals—both civilians and military personnel stationed at the fort. The "Slick Chick" column, for example, included a photo and brief "get acquainted" interview with a female civilian employee, while a "Soldier of the Week" column profiled an enlisted man.[12]

Though the Reception Station was the largest, most visible part of Fort Snelling's mission in World War II, there were several highly specialized military training units based there as well. Probably least known was the Military Railway Service, which for a time was headquartered at Fort Snelling. The Twin Cities, home to the Great Northern and Northern Pacific railway companies, had considerable local expertise to share. Working closely with civilian railroads, soldiers of the MRS literally learned how to run a railroad, from laying track to maintaining engines and rolling stock to dispatching trains. They also learned about foreign railroads, particularly in Europe,

Soldiers training on skis, 1942. *MNHS.*

and the MRS battalions were able to repair and operate railways in liberated territory to support and supply allied troops.[13]

Another specialized unit, the 99th Infantry Battalion, was organized specifically to take part in a possible Allied invasion of Norway. Recruits were required to have familiarity with snow-shoeing and skiing—and at least a conversational knowledge of Norwegian. Because Germany surrendered before the planned invasion of Norway, the 99th did not have to use the training they received in mountain combat in Colorado. But the unit served in Norway after the war, helping to process surrendered German soldiers on their way back to Germany.[14]

Possibly the most influential special unit at Fort Snelling was the most secret and thus least known at the time. The Military Intelligence Service Language School was made up of Nisei, American children of Japanese immigrants, primarily from the West Coast and Hawaii. Though many of these men had some background in the Japanese language and culture, the school sought to make them experts and introduce them to the particular words, terms, and idioms used by the Japanese army and navy. In November 1944, the male recruits were joined by fifty-one trainees from the Women's Army Corps.

The school was the brainchild of two army officers, Lieutenant Colonel John Weckerling and Captain Kai Rasmussen. Both men had been stationed in Japan prior to the war and recognized that in the event of a conflict between the United States and Japan, knowledge of the Japanese language would be critically important to American intelligence activities. After many delays, the Fourth Army Intelligence School, as it was initially titled, was opened at the army's base at the Presidio in San Francisco in November 1941, just a month before the attack of Pearl Harbor. Anticipating that few Anglo-American citizens would have the linguistic and cultural background to quickly become proficient in Japanese, the school recruited heavily among the Nisei population, many of whom had at least some exposure to the language, at home or in language schools, while growing up. Most of the instructors were Nisei as well, some of them academic scholars.[15]

On February 19, 1942, President Roosevelt ordered the removal from the West Coast of all people of Japanese descent—both immigrants, who were not eligible to become citizens and thus counted as "enemy aliens," and their American-born children, who were citizens. They were forced to abandon their homes and businesses and move to concentration camps set up in remote

Noboru Kinoshita at the chalkboard, Military Intelligence Service Language School, about 1945. *Photo courtesy of Karen Kinoshita.*

Among the sixty-one army officers and enlisted men from Minneapolis and the northwest who arrived at Fort Snelling in May 1945 after being liberated from German prisoner of war camps were (left to right): Staff Sergeant Carl W. Erickson, Staff Sergeant Robert Delange, Staff Sergeant Floyd Severson, Private First Class Alden Peterson, and Staff Sergeant Arnold W. Melin. *Lawrence Fuller Fort Snelling Research Materials*, MNHS.

areas around the western United States. As a result of this policy, the language school had to find a new home. While several states declined its request to relocate, Governor Harold Stassen of Minnesota accepted, and the Military Intelligence Service Language School, as it was now called, relocated to Camp Savage, about twelve miles up the Minnesota River from the confluence, near the town of Savage. The site had formerly been a state home for indigent men, and the location provided a degree of secrecy for the MISLS's classified mission. The Minnesota location also appealed to Captain Rasmussen because of its close proximity to Fort Snelling and the relative acceptance his Japanese American students would find in the Twin Cities.[16]

The school continued to grow, recruiting students and teachers from the camps—people who volunteered in spite of the federal government's unconstitutional treatment of their families and communities. It eventually outgrew this post and was moved in 1944 to Fort Snelling, where it became an important and integral part of life at the fort. The students were regularly featured in the post newspaper, as the school's formerly classified

nature was no longer a concern. The MISLS trained more than three thousand linguists and for a time was a major presence at Fort Snelling, utilizing more than 125 classrooms with a faculty of 162 instructors. A Chinese curriculum was added to the school in December 1944, and a small class of trainees in Korean was added in October 1945.

The work of the school graduates proved invaluable and may well have saved the lives of thousands of American soldiers. Following the war, MISLS graduates were vital to occupation forces, translating conversations and documents and providing important cultural insights. The school moved to Monterey, California, in June 1946.[17]

In September 1944, the processes at Reception Station No. 16 began working in the other direction. Men from the region returning from overseas reported to the station, where they received back pay, ration stamps, any needed clothing, and a physical exam. And then, within twelve hours of arrival, they were sent on their way home. At first

Soldiers lower the garrison's flag for the last time as Fort Snelling is decommissioned and turned over to the Veterans Administration. *MNHS.*

these were mostly GIs on extended leave or furlough, some of whom had been prisoners of war. A much larger-scale process began in 1945–46 in the Fort Snelling Separation Center, which processed men whose enlistments were over. These functions continued at the post until the army transferred them to Camp McCoy, Wisconsin, in July 1945.[18]

The move of the separation center and the language school from Fort Snelling confirmed what some in the wider community had already begun to suspect: its days as an active military post were coming to an end. The lack of any new, permanent construction indicated to some local leaders that the War Department did not see much of a future for Fort Snelling. On June 21, 1946, the last issue of the *Fort Snelling Bulletin* carried a message from Post Commander Colonel Harry J. Keeley confirming that the post would be closing, though he gave no specific date. He described how the

war had changed the position of the United States in the world and the place of the armed forces in American foreign policy. The country needed a permanent, large peacetime army, something unprecedented in US history. There was no longer a place for a relatively small post like Fort Snelling, with no room for expansion. "Military posts of the future peace time Army must accommodate about 30,000 men to provide for Divisional and Combined Arms training. So Fort Snelling must go," Keeley explained.[19]

As debate began over the future of the site, the military functions at the post were shutting down. On October 15, soldiers of the War Department Personnel Center lowered the garrison flag in front of the headquarters building for the final time. Only some 234 soldiers of the headquarters detachment remained to pack up their offices and shut down the building. The military career of Fort Snelling that had begun as the colonial outpost of a struggling republic ended in an urban center of a global superpower.[20]

Near Fort Snelling, about 1955. *Oil by Clement B. Haupers, MNHS; gift of Dr. John E. Larkin Jr.*

14. *The Making of Historic Fort Snelling*

IN THE DECADE FOLLOWING WORLD WAR II, officials from the military, the federal government, and the Twin Cities debated the future of the 6.7-square-mile remnant of the old Fort Snelling Reserve, first claimed from the Dakota under Pike's treaty. The area, formally known as Fort Snelling Unorganized Territory, continued to be important for the Twin Cities region and the federal government. And as it had since 1837, the large expanse of real estate in a rapidly growing region attracted interest and frustration.

On the one hand, the federal Department of Veterans Affairs (VA) was advancing its plans for the property. With a large number of veterans in the region, both the aging World War I veterans and many thousands more recently demobilized from World War II, the VA faced an enormous task. Across the country, there were nearly 20 million veterans, most of whom qualified for medical assistance, educational opportunities, housing loans, disability aid, and other benefits. In June, the Twin Cities VA branch took over the old riding hall, which saw yet another conversion, from dance and recreation center to a giant office space. In August 1946, the VA operations already

underway there required a workforce of over two thousand and were utilizing much of the space at Fort Snelling which the army had turned over to the agency when the post ceased to act as a military base. The American Legion and the Veterans of Foreign Wars, politically powerful veterans organizations, were unanimous in their support of the VA.[1]

On the other hand, the Minneapolis–St. Paul Metropolitan Airports Commission (MAC), which initially included the dynamic young mayor of Minneapolis, Hubert H. Humphrey, also wanted much of the Fort Snelling property. The war had brought great advances in aviation, and the potential for commercial air travel had important implications for the Twin Cities region. As early as December 1945, the MAC had identified Wold-Chamberlain Field, which was then a part of Minneapolis and Richfield, as the preferred site for development as the region's primary passenger terminal. To become a major regional airport, the facilities would have to expand dramatically, and the commission believed Fort Snelling the logical site for such an expansion. In a December 1945 press release, the MAC argued

Early Candlelight

Many Minnesotans' views of the history of Fort Snelling were shaped by the historical novel *Early Candlelight*, written in 1929 by Minnesota author Maud Hart Lovelace. The romantic tale was set in the 1830s at Fort Snelling, and it included depictions of real historical figures like Lawrence Taliaferro, Dr. John Emerson, Vital Guerin, and a main character, Jasper Page, loosely based on Henry Sibley. As historian Rhoda R. Gilman pointed out, however, "Like all other historical fiction, *Early Candlelight* is a double mirror. It reflects not only the period in which it is set, but also the times in which it is written. . . . This can distort the image and conclusions about people and events of the past that today are seen from a wholly different angle." Based on official records, letters, and diaries from white soldiers, officials, and civilians, the book necessarily reflects those viewpoints. It also embraced the beliefs current in Lovelace's day in American exceptionalism and the inevitable progress of American civilization, regardless of cost to Native peoples or the environment. At the time, the book was popular and well received. The ever-astute Colonel Walter C. Sweeney embraced the publication and its author, honoring Lovelace with a formal review of the troops at the post (supposedly the first time a female was so honored) as well as a gala reception for her and her guests at the officers' club.[i]

the need to acquire the greater part of the Upper Post, defining the question as one "weighing the importance of the continuation of Fort Snelling as a military post" against the need for 3,100 acres of land for what they saw as a critical expansion of air service. The public announcement of this proposal created an intense backlash, particularly from the influential veterans associations, who favored the VA.[2]

Thus an impasse developed between the MAC and the VA. The local agency, which had no direct authority over this federal property, appealed to the federal Civil Aeronautics Administration to intervene on its behalf with the VA. The Minnesota congressional delegation was also involved, seeking a compromise that would both support the Veterans Affairs Department and allow airport expansion. When the VA maintained its position at Fort Snelling, the MAC revised its plans. In 1948 and 1949, it began to acquire land in Richfield and Minneapolis, substantially reducing the amount of Fort Snelling reserve lands required for the expansion project. In an agreement reached in February 1953, the VA deeded 450 acres of Fort Snelling property to the MAC, and during the following year, the MAC came to an agreement with the air force over its needs. In 1956, an additional 250 acres of Fort Snelling lands were turned over for airport development.[3]

In 1951, the air force, by then an independent branch of the US military, proposed locating a new training facility for radio operators on the property, hoping to save money by reusing some of the structures not occupied by the VA. But that plan would have removed a runway from Minneapolis–St. Paul International Airport (it was renamed in 1948), so the idea quickly died.[4]

Yet these proposals were concerned mostly with the Upper Post, the area of the Fort Snelling Reserve developed between 1870 and 1942. The original structures built by Josiah Snelling's 5th Infantry and its successors prior to the Civil War were largely ignored. The outer walls of the old fort had been demolished before 1900. Except for a brief period in the early 1900s, when there had been an effort to refit the buildings, many of the original structures gradually fell into disuse, and most were torn down. By the 1950s, only the commandant's house, the officers' quarters, the hexagonal tower, and the round tower remained. Both the commandant's house and the officers' quarters had been largely rebuilt and greatly altered in 1905, with little of the original structures remaining.[5]

The most prominent of the surviving buildings was the round tower. It was a distinctive feature of the original fort, and over the years it had become a touchstone for memories of the region's "pioneer" colonial past. In 1939, the tower in fact became something of a shrine when it was remodeled as a museum, complete with military and Native American artifacts and a mural celebrating, as the *Minneapolis Tribune* put it, "the fort's role in military and civilian affairs." Much of the project was funded through private contributions, but the WPA carried out the work of renovation. Charles Richard Haines painted the mural as part of the WPA's Federal Art Project. The Minnesota Historical Society (MNHS) was also active in proposing the creation of the Round Tower Museum and ran it in partnership with the army. The post provided maintenance and custodial

Major Floyd Eller, commandant of Fort Snelling, and artist Charles Richard Haines in the round tower, May 1941. *Lawrence Fuller Fort Snelling Research Materials, MNHS.*

services for the building, while the historical society created appropriate exhibits. When the Veterans Affairs Department took over the fort in 1946, this arrangement fell apart, and the round tower was closed until 1965.[6]

The great expansion of buildings, roads, and other infrastructure during World War II overshadowed this remnant of the original post. The area around the original fort and the confluence was a backwater, cut off by roads and trolley lines. And when Fort Snelling ceased to be an active post, the round tower became something of an orphan. Though various local entities had an interest in the area, it was not a priority for the federal authorities who faced more pressing demands. From 1946 through 1948, MNHS continued to be concerned about its eventual fate. Proposals were put forward to have the area around the original Fort Snelling designated a national monument, but the National Park Service did not believe it had real national significance. There was also talk of making the area of the old fort and the confluence a state park, but despite these efforts, little headway was made.[7]

Continued development of the area, the enlargement of the airport, and growing suburbs meant that questions regarding the site could not be delayed forever. In 1956, the Minnesota Department of Transportation proposed changes in local highways that would have destroyed remnants of the old fort.

The director of MNHS, Russell W. Fridley, was relatively new to the position and to the state of Minnesota. A native of Iowa with a master's degree in history from Columbia University, Fridley was only twenty-seven when he was named director in 1955. But he was a hard worker and politically astute, with a natural talent for connecting with people. He recognized that the plan, while a threat to the historic site, was also an opportunity to gain public support for its preservation. As Minnesota Highway Department Director Michael Hoffman,

whom Fridley later described as "old school," proposed it, the route would run west on Fort Road (West Seventh Street), crossing the Mississippi and meeting another highway just north of the round tower—requiring a cloverleaf connection. The round tower would be surrounded by a loop of this intersection, completely cut off from public access. "And it was that cloverleaf around the Round Tower," Fridley later recalled in an interview, "that I think really caught the imagination of the public. Because in the minds of most people—including myself—I think the Round Tower *was* Fort Snelling."[8]

At a public meeting held in May 1956, the various stakeholders—including historical organizations, the highway department, the Minneapolis Parks Department, the MAC, the VA, and others, but apparently no one from the Dakota or Ojibwe communities—discussed the future of the old Fort Snelling site. The parties agreed that the area should be preserved "in an adequate setting," Fridley remembered. There was also a great deal of public support for the preservation of the site, "A spontaneous stream of telephone calls, letters and personal visits from patriotic and veteran's organizations, businessmen, labor leaders, teachers, school children and numerous individuals in sufficient number to demonstrate the widespread interest throughout the Northwest in the old fort site."[9]

In response to the public and private interest in the matter, Governor Orville Freeman asked the highway department, the conservation department, the historical society, and other interested agencies to come up with another plan. Fridley later recalled that the negotiations were contentious at times, particularly with the highway department and its deputy director, Frank Marzitelli. Once they reached a compromise, agreeing that the new highway should be routed through a tunnel, Marzitelli insisted that the tunnel be shortened, perhaps to reduce costs.

Freeway construction, 1961, showing the isolation of the round tower. The road cutting through the center of the diamond was moved and covered. *Lawrence Fuller Fort Snelling Research Materials, MNHS.*

The resulting cloverleaf cut off the Fort Snelling Chapel from the rest of the site. Yet the overall objective of saving the location of the old fort had been achieved.[10]

Contributing to the interest in the early fort and adding further impetus to the idea of its preservation was the impending state centennial celebration in 1958. During 1957, MNHS worked closely with the Minnesota Statehood Centennial Commission on planning events to commemorate the state's founding. The commission also provided $25,000 in funding for the society to sponsor an archaeological survey of the old Fort Snelling site during 1958. The excavations soon revealed that foundations of the outer walls, barracks, magazine, and other important structures

remained intact. They also uncovered many artifacts from the whole range of the fort's existence, including the years of its construction.[11]

To historical society staff, this was exciting information, suggesting possibilities for presenting Fort Snelling's history beyond the few exhibits at the round tower. In a 1958 presentation, Fridley described old Fort Snelling as "most central to the development of Minnesota" and being "remarkable that so much endures of a frontier fortress located near a large metropolitan area." The perspective at this time was focused on the colonial-era history of the site; the long history of the area's significance to the Dakota was not taken into consideration. He suggested that its woodland setting near the junction of two "great rivers" made it an ideal location for a "historical park." He presented two possibilities. The best way to take advantage of this "rare opportunity," he argued, would be a complete restoration of the fort as it stood when completed in 1824. Fridley recognized this would be a costly project, but in his view, it was well worth it. A less-expensive option would be to continue to operate the round tower and to complete the archaeology on the site, exposing the foundations of all the buildings "plus the removal of the obscuring second-growth trees and other modern distractions," thus restoring the "historic atmosphere of the area at the confluence of the two rivers." He offered the second option as the most easily attainable, while seeking funding for the more ambitious vision of a total restoration.[12]

The work on Fort Snelling was unprecedented for MNHS, which had a long-standing aversion to owning or managing historical properties. This stance was changing in the late 1950s, as the society acquired the William G. LeDuc House, its first historic site. Other historic properties within the state, notably the Charles A. Lindbergh House in Little Falls, were owned and maintained by the Minnesota Parks Department, while the

Fort Snelling State Park

When the National Park Service declined to regard the area around Fort Snelling as a site of national significance, the State of Minnesota began considering how the area might be used as a state park. In 1960, a study by landscape architect A. R. Nichols determined that the area along the river banks should be left in its "natural state" while some other areas in the former military reserve might be developed for recreational purposes. In late 1960, the Veterans Administration designated 320 acres, including the historical fort site and its immediate surroundings, "surplus property." This meant that the land could be transferred to the state without cost.

At about this time, the Fort Snelling State Park Association was formed to lobby the legislature to further develop the park and acquire additional lands. The association's board included Russell Fridley, director of MNHS; Udert Heller, director of state parks; Samuel H. Morgan, founder of the Minnesota Council of State Parks; and local businessman Thomas C. Savage. The park was dedicated in 1962, and during the 1960s and 1970s, additional lands acquired from government and private sources expanded the park to nearly 1,700 acres, including Pike Island and the site of the Civil War-era concentration camp. Though the initial focus of the park was on recreation and natural history, the cultural significance of these sites to the Dakota people is now recognized and commemorated in the park.[ii]

Minnesota Historical Society provided interpretive information and exhibits. For a time, this arrangement worked well enough, but problems arose during the 1960s. First, the sites were not a priority for the parks department; in Fridley's view, the state parks folks were not maintaining them properly. Second, from the legislature's point of view, funding went with the agency that actually held the property. So while sites were languishing or deteriorating, MNHS could not obtain funding to preserve or develop them. This situation led Fridley to reconsider the society's past reluctance to own and manage historic sites.[13]

This was the context in which the development of Historic Fort Snelling began, and it marked an important transition in the institutional history of the Minnesota Historical Society. Though the historical society had received significant state support, growing from an office in the state capital to having its own building occupying a prominent place on the Minnesota State Capitol grounds, its focus into the 1950s had been on collecting and publishing. Creating a reconstruction of old Fort Snelling would require significant government

Conservationist Sigurd Olson joined Thomas Savage and Russell Fridley at the new state park, about 1962. *MNHS.*

funding and the necessarily public support to go with it. The Minnesota Historical Society would have to become more visible to the public beyond St. Paul, and developing more historic sites around Minnesota proved an effective means of accomplishing this goal.[14]

There was significant political support for the society's ambitions. Early on, the most important backer was Governor Elmer L. Andersen. A liberal Republican and progressive business leader of the H. B. Fuller Company, Andersen was a champion of public education and a lover of history. He proved to be an important early advocate for the Fort Snelling project and the society's historic preservation goals. Governor Andersen had been a member of the society's executive council in the 1950s, so he knew the institution well and provided important support for its budget requests. There were key connections within the legislature as well, many of them from greater Minnesota. These included Richard Fitzsimons of Argyle, chair of the house appropriations committee, and Gordon Rosenmeier of Little Falls, one of the most influential members of the senate. Both had a great interest in historic preservation and sites within their districts. Together with Andersen, they helped craft the society's 1963 appropriation, which still went through the Conservation Department but garnered $80,000 to produce three Minnesota Outdoor Recreation and Resources Commission (MORRC) reports on historic sites, archaeology, and Fort Snelling.

Two years later, the reports were completed, and MORRC approached MNHS for a site that could be a "showcase" for its work. The Fort Snelling reconstruction was the clear choice at that time. The project received a grant of $100,000 from the MORRC, a large sum in 1965, which attracted further grants and appropriations from state, federal, and private sources over the next decade, amounting "into the millions," as Fridley remembered, to complete the restoration.

Beginning with the heightened interest in Fort Snelling that arose from the statehood centennial and the site's near destruction, propelled by helpful legislative leadership and support from Minnesota governors, the rebirth of the old fort in the form of Historic Fort Snelling came to be.[15]

⌒

But what would this resurrected nineteenth-century edifice, "Historic Fort Snelling," look like? What would it mean for the people who visited it and the community around it? An accurate reconstruction of the structures required extensive archaeological work augmented by historical research along with carefully considered rebuilding techniques. How to present the fort's history and the lives of those who occupied the buildings proved a more complex and challenging task.

Early on, the decision was made to restore the fort to its appearance during the period between 1824 and 1848, identified by planning staff as Fort Snelling's "peak years," and to concentrate these efforts on the buildings within the original diamond-shaped post. In part, this choice was dictated by the archaeology, since the stone walls of that period had shown up so clearly in the excavations. There were also plenty of written sources, since the army's bureaucracy had created accounts, reports, maps, even drawings and plans useful for any reconstruction of the early nineteenth-century structures. Most of these records were available through the National Archives. These same sources, along with written accounts and records of individuals like Lawrence Taliaferro, Philander Prescott, Charlotte Van Cleve, the Pond brothers, and others, could provide rich details for interpreting the lives of soldiers, missionaries, and civilians at the post. The fact that these sources would dictate a strongly white, military perspective, focused on official life within the limestone walls, seems to have mattered little in the early 1960s. Though the growing civil rights

Excavating in the round tower in preparation for restoration, August 1965. *Photo by Terry Garvey, MNHS.*

movement was in the daily headlines, it had yet to cause a reexamination of American history or the consideration of colonialism and racism in the nation's growth and expansion.[16]

The reconstruction took the better part of a decade of careful work. The society reopened the Round Tower Museum in the summer of 1965, with a curator and guide available for members of the public who made their way to the site. A second archaeological dig, seeking further details on the original structures at the fort, began at the same time. These and later excavations uncovered several surprises, such as remnants of the original officers' barracks beneath the 1905 renovations and underground storage tunnels running from their basements under the parade ground. Many thousands of artifacts were found on the

site, including a six-thousand-year-old projectile point, showing that Indigenous peoples have been at Bdote for thousands of years.[17]

Builders went to great lengths to use techniques and materials from the 1820s, even reproducing period tools when none were available and training craftspeople in their use. Much of the hardware, including hinges, pintles, latches, and nails, were manufactured by blacksmiths working on the site, though they used a form of soft steel rather than less durable wrought iron. Period window glass was manufactured by a specialty glass company in West Virginia using late eighteenth-century techniques.[18]

Since the resurrected fort would see many modern visitors, planners made accommodations to ensure comfort and safety. The original fort used lead paints and whitewash on many of its walls and outer surfaces. Whitewash could be troublesome, since it tended to rub off on people's clothing, so it was replaced with flat white latex paint applied over a specially prepared surface so that it resembled whitewash. The original outer walls were rebuilt on a base of reinforced concrete with a vertical core that was then faced with limestone, thus making them much more stable than the original stone walls. The schoolhouse was rebuilt with concealed electrical fixtures, since it would be used for public programs, and all of the exposed wooden structures were made of wood treated with preservatives for added durability.[19]

The result of this painstaking effort was a version of Fort Snelling circa 1824, authentic in appearance, which opened to the public in phases beginning in 1967 and in fully realized form in 1977. To enhance the experience for visitors and to provide additional educational value, trained staff, wearing clothing appropriate to the 1820s, explained and demonstrated life in a frontier post. Starting in 1970, a small group of young guides, some of them cadets from St. Thomas Academy, a local military high school, were hired

Window glass made with eighteenth-century techniques
in the re-created officers' quarters. *Photo by Matt Schmitt.*

The parade ground of the reconstructed fort, 2016. *Photo
by Matt Schmitt.*

Interpreter training, 1976. *Photo by John Grossman, MNHS.*

to act as soldiers at the fort, wearing uniforms from the 1820s and depicting narrowly scripted characters. As the site was developed, the number of interpreters and their roles expanded but remained closely focused on the military history of Fort Snelling. Site guides even developed their own military-style organization, a "Fort Snelling Guard," with more experienced guides acting as noncommissioned officers supervising less-experienced staff.[20]

By the completion of construction at Historic Fort Snelling, a carefully researched program was in place. Interpreters used a living history "first-person" style, modeled on similar programs at Colonial Williamsburg and Plymouth Plantation, in which they portrayed real, historical individuals and spoke to visitors as if it were 1827. The program aimed to create an experience both educational and entertaining. Though still centered within the fort's walls, the interpretation widened somewhat to include women, both officers' wives and laundresses, as well as Indian agency employees and Selkirkers living on the military reservation.

By 1976, as the United States celebrated the bicentennial of the Declaration of Independence, Historic Fort Snelling presented a traditional historical narrative, celebrating the outpost of American expansionism with military pomp and circumstance. Public reaction was positive. So adept were the interpreters at their role-playing that they occasionally led visitors to suspend disbelief—or to play along. One family offered groceries to "starving" soldiers and another couple wanted to adopt a soldier they thought was too young to be in the army. Historic Fort Snelling soon became, and remains, among the Minnesota Historical Society's most frequently visited historic sites.[21]

But there were drawbacks as well. First-person interpretation required that interpreters receive hours of training and testing on the social and historical background of the period. The system worked best if interpreters stayed on for a number of years, which was not always possible with seasonal staff earning entry-level wages. First-person interpretation also asked visitors to play along with the fiction of 1827, which occasionally made them uncomfortable. Some hesitated to interact, fearing they might say something "wrong," and on a few occasions, staff reinforced these concerns—by going too far in correcting visitors' manners, for example.

Setting the re-created fort in the year 1827 also constrained the interpretation of early Minnesota history presented at Fort Snelling. How would issues like slavery, particularly the story of Dred and Harriet Scott, be told in a first-person setting in the year 1827, long before they arrived at the fort? In addition, the year was not a particularly happy one, with killings and executions of Ojibwe and Dakota people as well as dissension among army officers and the hasty departure of Colonel Snelling. The broader culture was also

Interpreters in fife and drum demonstration, 1973. *Photo by Monroe Killy, MNHS.*

shifting: civil rights activism in the 1960s and 1970s inspired the founding of the American Indian Movement in Minneapolis, and some of the fort's visitors were arriving with different questions about the past.[22]

The site's staff—predominantly white, with few people of color—and the historians they worked with were well aware of these limitations. The mostly empty buildings of the Upper Post showed that there was much more to the overall story of Fort Snelling than what took place inside the rebuilt diamond, and they developed additional techniques to make the public aware of it. Information offered by interpretive staff, exhibits in the officers' quarters, a teacher's guide discussing such issues as Manifest Destiny and Indian-white relations, special events like Civil War Weekends—all attempted to broaden and add context to the visitors' experience. But as one staffer noted in 1980, the military tone set in the 1970s had created an image "that has been both immensely popular and sometimes annoyingly limited to those who realize the remarkable potential this site has to offer." The military tone was not popular, however, among many people of color and Indigenous people who did not see their history told well.[23]

The opening of the Historic Fort Snelling Visitor Center in 1982 offered an opportunity to present a more complex version of the post's history. The center housed small exhibits, including one on the 25th Infantry. After watching an introductory film, visitors walked from the center to the fort and received a brief orientation, then took self-guided tours, interacting with interpreters representing various roles. By the late 1980s, on a busy summer day, about twenty-five guides in costume presented the experiences of soldiers, Indigenous people, laundresses, fur trade employees, officers' wives, and Selkirkers in 1827. Interpretation broadened to include more information on the Indian Agency, the immigrant experience among soldiers and their dependents, and nineteenth-century medicine. Interpretive staff began to adopt a more flexible style, dropping character to answer visitors' questions and generally making the public feel comfortable and welcomed.[24]

For many interpreters, the job represented more than a paycheck, modest as it was. Their love of history and devotion to their work led them to commit considerable time and effort beyond their paid hours in educating themselves in the details of daily life in the early nineteenth century, studying interpretive programs at other institutions, and developing ways to apply what they had learned to their work at Historic Fort Snelling.[25]

⁓

In the late 1990s, MNHS considered refocusing the interpretation at Fort Snelling to 1838, a year with a broader array of stories: the 1837 treaties, slavery at the fort (particularly the lives of Dred and Harriet Scott), the decline of the fur trade, and the interactions between the garrison and the wider community around it. The proposal was not adopted, evidently because of costs, although these efforts certainly had some influence on programming. The emphasis remained on interpreting life inside a US military post during the first half of the nineteenth century. The scope of interpretation continued to expand, with an exhibit on the Military Intelligence Service Language School and the lives of Japanese Americans during World War II. Another exhibit entitled "History Under the Floorboards" focused on archaeology but included information on the lives of enslaved African American "servants" at Fort Snelling; however, it did not directly address the larger issue of slavery at the fort.[26]

From its beginnings, the historic interpretation at Historic Fort Snelling had been driven by two factors: the institution's desire to teach the public about aspects of local and American history they may not have known, and the input that the staff received from the public, including visitors' questions and comments on diverse aspects of the life in the past as it was presented at the fort. These often reflected the dramatic social changes in the late twentieth century; the sharing of bunks by soldiers, for example, began to draw questions about sex and sexuality that were less likely to be raised in the 1970s. Interpretive staff members from that era remember a continual move to modified first-person interpretation through the 1990s and early 2000s. Interpreters shifted to demonstrating or explaining aspects and issues of life at Fort Snelling over the years, rather than discussing their "lives," which allowed for more flexibility in content and freer interaction with the public. Interpreters could discuss slavery at the fort, for example. In 2007, first-person interpretation was formally ended.[27]

A further public influence on interpretation at Fort Snelling arose from communities that had been largely excluded or ignored at the time of Historic Fort Snelling's original planning. African American organizations, particularly in the Twin Cities, urged MNHS to continue to recognize the presence of Dred and Harriet Scott and to expand its examination of slavery at Fort Snelling. This encouragement spurred further development of interpretation, which has been greatly assisted by recent scholarship revealing the army's role in spreading and maintaining slavery in the Northwest, particularly at Fort Snelling.

Indigenous people have been most intensely critical of interpretation at the fort, and rightly so. A small number of Dakota staff members struggled over the years to have their voices heard. Many Dakota people (including tribal leaders and scholars) have pointed to shortcomings in the presentation of Dakota history, arguing for a thorough treatment of the fort's impact on their relatives, past and present, and the ways the site represents white colonization and genocide against the Dakota people. In 2002, organizers

began the Dakota Commemorative March, a biannual 150-mile walk in November from Camp Release to the site of the concentration camp below the fort. In May 2010, a number of Dakota activists and allies carried out a "Take Down the Fort" demonstration and protested on the fort's parade ground.

In 2016, MNHS sought funding for a new museum and interpretive center at Historic Fort Snelling from the Minnesota legislature. This request, which would allow programing to include greatly expanded stories of Dakota people at Bdote, enslaved people at the fort, and the Japanese Americans at the language school, met with a very different public reaction than the 1960s proposals for the original reconstruction. Dakota protestors demonstrated against funding for the fort, and some again called for the deconstruction of Historic Fort Snelling; other Minnesotans, objecting to the expansion of the interpretation, also testified against the funding. In the words of Dakota scholar Waziyatawiŋ (Dr. Angela Cavender Wilson), Historic Fort Snelling had already become "a metaphorical representation of the ongoing celebration of colonialism in Minnesota."[28]

Other people in the Dakota community have been reclaiming their historical place in the local landscape by other means. Their efforts led to the restoration of the Dakota name Bde Maka Ska to Lake Calhoun, for example. And through programming sponsored by the Minnesota Humanities Center, Dakota educators are promoting a broader recognition among other Minnesotans of Bdote as an area sacred to Dakota people.

Administrators and staff at MNHS had been working for decades to improve relationships with Native communities. In 1987, MNHS established an Indian Advisory Committee, which provides guidance on programming. Before and during the sesquicentennial observation of the US–Dakota War in 2012, Dakota people held four Dakota

Historic Fort Snelling from across the Mississippi, 2016.
Photo by Matt Schmitt.

Iyuha Owanjina (Together As One) Nationwide Conferences, in which descendants of the Dakota diaspora from Minnesota, Nebraska, North and South Dakota, Montana, and Canada reunited to observe the anniversary and to teach and learn their history. MNHS provided meeting space and other support, and its staff members were allowed to attend many sessions. The meetings were powerful and painful. Attendees visited sacred sites throughout Bdote; many entered the reconstructed fort for the first time. In a separate set of meetings, MNHS exhibits staff sought advice on how to handle deeply problematic objects in the institution's collections. Many MNHS staff members who participated found the experience to be transformative.[29]

The creation at MNHS of a Native American Initiatives department in 2017, and an ongoing partnership with community leaders called the Dakota Community Council (DCC), has more recently helped advance more inclusive interpre-

tation and has led to more Dakota community participation in decisions made at the site. This partnership and engagement work encouraged more robust natural landscape preservation in this space, for example. Though not without barriers and setbacks, this difficult but necessary work continues.

In 2017, MNHS altered the sign at the entrance to the twenty-three-acre Historic Fort Snelling site with a temporary covering that read, "Historic Fort Snelling at Bdote," recognizing both the fort's original Dakota context and the ongoing significance of the confluence. Interpretation at the fort concentrated on the stories of Bdote and the Dakota, military history, slavery and freedom, and the World War II Japanese language school. While the site's military history continued to be the primary focus of its interpretation, the publicity around the new interpretive programs led to a reaction from some in the white community. They feared that the new emphasis would diminish the

telling of the military history of Fort Snelling, if not obscure it entirely. In addition, changing the name of a historic site requires an act of Minnesota's legislature. Protests arose. In spring 2019, MNHS removed the temporary signage and began a process for considering a new name.[30]

This deep interest in the site and its interpretation is not surprising. History matters. The new Fort Snelling Visitor Center, which will continue to tell the complicated, contentious, important, and ever-evolving story of the people at the confluence, is scheduled to open in 2022.

It is interesting to consider how history might be understood at the confluence if MNHS had adopted Russell Fridley's second option regarding the fort's preservation: a restored round tower surrounded by ruins. The site would be much more open to interpretation and connected to landscapes both old and new: Bdote, the footprint of the old limestone fort, and the Upper Post. The meticulous reconstruction of the post as it stood (or as it was thought to have stood) in 1827 locked the story into a largely military perspective, and for many years that view focused mostly on the early nineteenth century.[31]

Despite this built-in bias, MNHS staff at Historic Fort Snelling have worked hard to tell the complex, sometimes tragic story of the fort and early Minnesota, a story shaped by the broader context of American history. The telling of that story has changed since the 1960s, becoming more inclusive as new scholarship and social attitudes have opened a wider view of the nation's history, and it will continue to evolve. It is a story that needs to be understood. After all, we live in the world created by the policies and ambitions that built Fort Snelling. For good and for ill, the historical currents that formed that place still affect our lives.

Acknowledgments

RESEARCHING AND WRITING THIS HISTORY OF Fort Snelling has proven a challenging and exciting project. It is a complex story, reflecting many of the themes of American history as the post evolved from a colonial outpost to a permanent military base of a great power. It is also part of a much older history, that of the Indigenous people of the region of Bdote, particularly the Dakota. For them Fort Snelling was a source of pain and despair, the center and symbol of the loss of a beloved homeland.

In writing this book I tried to focus on Fort Snelling and the people in and around it while making the connection between the post and the wider context of American and sometimes world history. Navigating these themes was tricky at times, and I benefited greatly from the input of the editorial staff at the Minnesota Historical Society Press, particularly editor Ann Regan. Our many conversations on events and issues covered here were particularly helpful. They added greatly to the focus and coherence of this work, when I might otherwise have wandered too far into the weeds.

My thanks to a number of individuals who graciously contributed their knowledge and expertise to the making of this book; to readers of early drafts, including Matt Cassady, Peter DeCarlo, and Kate Beane, who provided useful insights and corrections; and especially to staff members at Historic Fort Snelling, including site manager Nancy Cass and program specialist Jeff Boorom, who provided valuable perspectives into current and past approaches to historical interpretation at the fort.

My gratitude also to my former colleagues in the Minnesota Historical Society library, who provided their usual informed and professional assistance. Library technology keeps advancing, and sometimes even an old hand needs help navigating catalogs.

Finally, I would be more than remiss if I did not acknowledge the help of my spouse, Sally Reynolds. Her keen eye for corrections, ability to ask the obvious questions, and above all patience with the author's frustrations and anxieties were invaluable.

Notes

Notes to Prologue

1. Westerman and White, *Mni Sota Makoce*, 15.
2. See Osman, *Fort Snelling and the Civil War*, and Osman's series of short monographs, *Fort Snelling Then and Now*, which includes his survey of the fort's role in World War II, "Fort Snelling's Last War."

Notes to Chapter 1:
A Land and Its People

1. Westerman and White, *Mni Sota Makoce*, 85-111.
2. Westerman and White, *Mni Sota Makoce*, 15; Anderson, *Kinsmen of Another Kind*, 11.
3. Wingerd, *North Country*, 14.
4. Westerman and White, *Mni Sota Makoce*, 19, 29. The Ojibwe were then known as the Chippewa.
5. Pringle, "The First Americans"; Wells, *The Journey of Man*, 139-40.
6. Westerman and White, *Mni Sota Makoce*, 29.
7. Westerman and White, *Mni Sota Makoce*, 20, 26.
8. Westerman and White, *Mni Sota Makoce*, 122-23.
9. Anderson, *Kinsmen of Another Kind*, 29. Trade guns of this period were flintlock muskets, with a mechanism that relied on a spark produced by a stone flint striking a metal blade to set off the gunpowder. Though mass produced, they were essentially handmade and prone to failures of one kind or another: "Flintlock," Wikipedia, https://en.wikipedia.org/wiki/Flintlock.
10. Wingerd, *North Country*, 21.
11. U.S. History: Pre-Columbian to the New Millennium, 8a. New France: https://www.ushistory.org/us/8a.asp.

Though much larger in area, the population of New France was approximately one-twentieth the size of the English colonies: Jacques Mathieu, "New France," The Canadian Encyclopedia, https://www.thecanadianencyclopedia.ca/en/article/new-france.
12. Wingerd, *North Country*, notes to plates 10, 11.
13. Wingerd, *North Country*, 23.
14. Wingerd, *North Country*, 12.
15. Wingerd, *North Country*, 20.
16. Wingerd, *North Country*, 40-42.
17. Wingerd, *North Country*, 39, 44.
18. Wingerd, *North Country*, 32-33.
19. Wingerd, *North Country*, 32-33.
20. Anderson, *Kinsmen of Another Kind*, 27.
21. "French and Indian War," Wikipedia, https://en.wikipedia.org/wiki/French_and_Indian_War; Wingerd, *North Country*, 47.
22. Wingerd, *North Country*, 56.
23. Wingerd, *North Country*, 55.
24. Wingerd, *North Country*, 55, 64-68.
25. Anderson, *Kinsmen of Another Kind*, 74-75.
26. Anderson, *Kinsmen of Another Kind*, 15, 74; Wingerd, *North Country*, 63.
27. Wingerd, *North Country*, 59, 62-63.
28. Boyer, ed., *The Oxford Companion to United States History*, 701.

Notes to Chapter 1 Sidebar, page 6

i.   Orsi, *Citizen Explorer*, 95-100.
ii. Fremling, *Immortal River*, 34-39.

iii. Fremling, *Immortal River*, 40-54; Bray, "Millions of Years in the Making."

iv. Fremling, *Immortal River*, 59.

v. Fremling, *Immortal River*, 61.

Notes to Chapter 2:
The Arrival of the Americans

1. The word *frontier* reflects a European American perspective but is a useful, nearly unavoidable shorthand in discussions of this era.

2. Prucha, *The Sword of the Republic*, 53, 153, 199, 319. Case, *The Relentless Business of Treaties*, discusses the motives of treaty signers, including legislators who were involved in land speculation.

3. Prucha, *The Sword of the Republic*, 60; Wingerd, *North Country*, 72.

4. Thomas Jefferson to William Henry Harrison, February 27, 1803, cited in Case, *The Relentless Business of Treaties*, 40-41; Prucha, *The Sword of the Republic*, 72.

5. Prucha, *The Sword of the Republic*, 75-76.

6. Orsi, *Citizen Explorer*, 63, 64.

7. Orsi, *Citizen Explorer*, 75.

8. Orsi, *Citizen Explorer*, 79.

9. Wingerd, *North Country*, 47; Westerman and White, *Mni Sota Makoce*, map, 122-23.

10. Folwell, *A History of Minnesota*, 1:93. These conflicting ideas of land ownership existed in White-Indian relations as far back as the earliest English settlements: see Cronon, *Changes in the Land*, 58, 65. Lawrence Taliaferro discusses misunderstandings as to the extent of the land involved: Taliaferro journal, September 7, 1830. It is also impossible to know exactly what Pike's translator, Joseph Renville, told the Dakota or how he explained their views to Pike.

11. Anderson, *Kinsmen of Another Kind*, 82. Jean Baptist Faribault had traded at Bdote the previous year, but none of the American sources seemed to notice. He may have been wintering with one of the local Mdewakanton villages: see Denial, "Pelagie Faribault's Island," 51.

12. Westerman and White, *Mni Sota Makoce*, 141.

13. Orsi, *Citizen Explorer*, 80, 83-84.

14. Prucha, *The Sword of the Republic*, 99.

15. Prucha, *The Sword of the Republic*, 118. Britain's Indian allies and fur trade interests had wanted to force the United States to recognize the area between the Ohio River and the Great Lakes as Indian territory and cede any interests in the Upper Mississippi region.

Notes to Chapter 3: Building a Fort

1. Jacob Brown to Secretary of War, February 5, 1818, Brown Papers.

2. John Calhoun to Jacob Brown, October 17, 1818, Brown Papers.

3. John Calhoun to Jacob Brown, August 15, 1819, and John Calhoun to Hezekiah Johnson, August 1819—both US War Department Secretary's Office, Letters Sent, Military Affairs.

4. Forsyth, "Fort Snelling: Col. Leavenworth's Expeditions," 140.

5. Forsyth, "Fort Snelling: Col. Leavenworth's Expeditions," 149.

6. Forsyth, "Fort Snelling: Col. Leavenworth's Expeditions," 151, 152.

7. Forsyth, "Fort Snelling: Col. Leavenworth's Expeditions," 159.

8. Forsyth, "Fort Snelling: Col. Leavenworth's Expeditions," 154. The location is on what is now Picnic Island in Fort Snelling State Park.

9. General Jacob Brown to Secretary of War William Crawford, May 31, 1815, Brown Papers. Henry Rowe Schoolcraft sketched the post in 1820: see "Geological Mineralogical Journal," Schoolcraft Papers, MNHS microfilm M296, reel 48, p. 50.

10. Forsyth, "Fort Snelling: Col. Leavenworth's Expeditions," 156.

11. Like many officers of that time, Josiah Vose came from a military family. His father, Joseph Vose, was commander of the 1st Massachusetts regiment of the Continental Army during the Revolution. Josiah was commissioned captain during the War of 1812 and rose to the rank of major. He survived the various army cuts following the war and eventually became colonel of the 4th Infantry in 1842. He died in 1845. Heitman, *Historical Register and Dictionary of the United States Army*, 990; *Boston Post*, October 8, 1845.

12. Josiah Vose to Henry Leavenworth, October 1819, included in the letter of Leavenworth to Major General Alexander Macomb, October 24, 1819, National Archives Record Group [hereafter, NARG] 107, Letters Received by the Secretary of War, Main Series. Robert McCabe began his army career during the war with Britain, joining the 5th Infantry in 1816. Though he does not seem to have had a formal education, his engineering skills were considerable and much admired by his fellow officers. He designed and supervised the building of the mills at St. Anthony Falls and oversaw much of the building of the fort itself.

13. Josiah Vose to Henry Leavenworth, October 1819, included in the letter of Leavenworth to Major General Alexander Macomb, October 24, 1819, NARG 107, Letters Received by the Secretary of War, Main Series.

14. Henry Leavenworth to Alexander Macomb, October 24, 1819, NARG 107, Letters Received by the Secretary of War, Main Series.

15. Returns and Rolls, 5th US Infantry, NARG 94, US War Department Adjutant General's Office.

16. John Calhoun to Henry Leavenworth, December 29, 1819, NARG 107, US War Department Secretary's Office, Letters Sent, Military Affairs, MNHS microfilm M152.

17. Folwell, *A History of Minnesota*, 1:137; Parker, ed., "The Recollections of Philander Prescott," 29, as quoted in Wingerd, *North Country*, 90. Philander Prescott was originally from New York and came to the post as an assistant to

the 5th Regiment's sutler, Mr. Devotion. When the latter's business failed, Prescott became a fur trader.

18. Prescott, "Autobiography and Reminiscences," 479; Westerman and White, *Mni Sota Makoce*, 92; Henry Leavenworth to Thomas Jesup, June 18, 1823, US War Department Quartermaster Generals Correspondence; John Calhoun to Jacob Brown, October 17, 1818, US War Department.

19. Folwell, *A History of Minnesota*, 1:144 and note; Denial, "Pelagie Faribault's Island," 51.

20. Luecke and Luecke, *Snelling: Minnesota's First First Family*, 11-58.

21. Returns and Rolls, 5th US Infantry, October 31, 1820, NARG 94, US War Department Adjutant General's Office; Josiah Snelling to John Calhoun, November 10, 1820, NARG 107, Letters Received by the Secretary of War, Main Series.

22. Henry Leavenworth to Thomas Jesup, June 18, 1823, Fort Snelling file, US War Department, Quartermaster Generals Correspondence.

23. Prescott, "Autobiography and Reminiscences," 479; Josiah Snelling to Thomas Jesup, August 16, 1824, Fort Snelling file, NARG 92, US War Department Quartermaster Generals Letter Received, MNHS microfilm M222.

24. US General Accounting Office, Third Auditor, 1822, files 10872 and 11467, NARG 217, Fort Snelling Research Papers, P333, box 6, MNHS; Registers of Enlistments in the US Army, 1798-1897, NARG 98, Records of the Adjutant General's Office, MNHS microfilm M191.

25. US General Accounting Office, Return of Items used, worn out, etc., September 1822, NARG 217, Fort Snelling Papers, box 6.

26. US General Accounting Office, Return of Items used, worn out, etc., Josiah Snelling to Peter Hagner, March 18, 1823, 4th Quarter 1822, receipts for horses and mules, NARG 217, Fort Snelling Papers, box 6.

27. Josiah Snelling to Thomas Jesup, August 16, 1824, NARG 92, Office of the Quartermaster General, Consolidated Correspondence, Fort Snelling.

28. Letters Sent by the Office of the Adjutant General, January 7, 1825, NARG 94, Records of the Adjutant General's Office.

29. Taliaferro journal, typescript, 26; Black Dog Chief to Lawrence Taliaferro, Taliaferro journal, September 8, 1830.

30. Taliaferro journal, Josiah Snelling to Thomas Jesup, July 27, 1823, expresses doubt that supplies requested will arrive, "the Mississippi being fordable in several places between this post and the lake."

31. General Accounting Office Returns, July 1822, NARG 217, Fort Snelling Papers, box 6; Josiah Snelling to George Gibson, May 16, 1823, US War Department Quartermaster Generals Correspondence.

32. Josiah Snelling to George Gibson, October 6, 1826, Commissary General of Subsistence; James Gale to Thomas Jesup, October 4, 1829, US War Department Quartermaster Generals Correspondence. On James Gale, see Heitman, *Historical Register and Dictionary of the United States Army*, 1:442-43.

The "blackbirds" were quite probably the common grackle, a bird native to North America that at times moves in large flocks and has a fondness for grain, especially corn: The Cornell Lab, All About Birds: Common Grackle, https://www.allaboutbirds.org/guide/Common_Grackle/overview.

33. Josiah Snelling to George Gibson, November 11, 1822, Commissary General of Subsistence.

34. Josiah Snelling to George Gibson, October 6, 1826, Commissary General of Subsistence.

35. Taliaferro journal, March 26, 1826.

36. Prucha, *The Sword of the Republic*, 186.

37. Taliaferro journal, October 23, 1827. The following evening Taliaferro recounted, "Tremendous fires on the prairies last night, light enough to pick up the smallest substance from the ground. . . . Public wood—20 cords which had been cut, corded up and paid for—was entirely consumed by the fires of the prairies yesterday evening." Given the events of that summer, discussed below, it is possible that the burning of his "public wood" may not have been entirely accidental.

38. See below, page 49; Taliaferro journal, December 6, 1826, January 5, 1828, July 8, 1831; Pond, *The Dakota or Sioux in Minnesota*, 151. Pond also noted that whites in general had a distorted view of Dakota women, thinking them generally "unchaste" when in fact most unmarried girls were shy and diffident, little different than in white society.

39. January 4, April 1, April 13, 1845, 1st Regiment Order Book; Electus Backus to Henry Turner, February 28, 1845—both Fort Snelling Papers, box 3.

40. Taliaferro journal, October 17, 1835, March 29, 1836; Hansen, *Old Fort Snelling*, 98; White and White, *Fort Snelling in 1838*, 120.

41. Men would not have thought this arrangement unusual, as people often shared beds in this era and it was standard army practice since it provided heat and saved space: Brown, *The Army Called It Home*, 71.

42. White and White, *Fort Snelling in 1838*, 98-100; Sutler's account books, MNHS microfilm M169, roll 28, Sibley Papers.

43. National Park Service, Fort Scott, "Laundress-Historic Background," https://www.nps.gov/fosc/learn/education/laundress5.htm.

44. Surgeons Quarterly Reports of Sick and Wounded, July-September, 1847, Fort Snelling, NARG 94, Records of the Adjutant General's Office, MNHS microfilm M224; Haines, "Estimated Life Tables for the United States," 1. This number averages in the high rate of infant mortality.

45. Surgeons Quarterly Reports of Sick and Wounded, July-September, 1847, Fort Snelling, NARG 94, Records of the Adjutant General's Office, MNHS microfilm M224; Dr. E. Purcell to Surgeon General Joseph Lovell, January 10, 1824, NARG 112, Office of the Surgeon General, selected letters received, Fort Snelling Papers, box 3.

46. Letter and report of Dr. J. Ponte Coulant McMahon to Surgeon General Joseph Lovell, July 1, 1827, NARG 112, Office of the Surgeon General, selected letters received, Fort Snelling Papers, box 3.

47. Army Enlistments, Surgeon George Turner to Surgeon General Thomas Lawson, January 30, 1841, NARG 112, Office of the Surgeon General, selected letters received, Fort Snelling Papers, box 3.

Note to Chapter 3 Sidebar, page 37

i. Josiah Snelling to Thomas Jesup, August 16, 1824, NARG 92, Office of the Quartermaster General, Consolidated Correspondence, Fort Snelling.

Notes to Chapter 4:
The Agency and Its Community

1. Thomas Jefferson to William Henry Harrison, February 27, 1803, cited in Case, *The Relentless Business of Treaties*, 40-41.

2. Prucha, *The Sword of the Republic*, 94.

3. Taliaferro Papers, MNHS microfilm M35, roll 1, April 2, 1819. Major Taliaferro never completely recovered from his malady and from time to time suffered relapses during his career as Indian agent.

4. Taliaferro, "Auto-biography," 188; Taliaferro Papers, MNHS microfilm M35, roll 1, April 2, 1819.

5. Lawrence Taliaferro to John C. Calhoun, Taliaferro journal, March 1, 1822; Lawrence Taliaferro journal, September 9, 1821; Wingerd, *North Country*, 83; Anderson, *Kinsmen of Another Kind*, 111-14.

6. Prucha, *The Sword of the Republic*, 322-25.

7. Taliaferro journal, January 5, 1828, February 19, 1826. In fact, John Russick was something of a troublemaker in general. He was court-martialed twice in 1826 for drunkenness in the Company F barracks and reduced to the ranks from corporal to private for being absent from a wood-cutting detail: September-October 1826, 5th Regiment Garrison Orderly Book, MNHS microfilm M239.

8. Gilman, "The Last Days of the Upper Mississippi Fur Trade," 124.

9. See, for example, Zachary Taylor's observations in 1830, quoted in Hamilton, "Zachary Taylor and Minnesota," 101.

10. For an example of the agent's entreaties, see Lawrence Taliaferro to John Bliss, July 19, 1834, Taliaferro Papers. Taliaferro journal, Agent's Accounts. It should be noted that Taliaferro's spelling of Indian names was strictly phonetic and varied frequently over the years.

11. Green, *A Peculiar Imbalance*, 4-5; Wingerd, *North Country*, 90.

12. Wingerd, *North Country*, 91, 126.

13. Obst, "Abigail Snelling," 100, 101. Josiah Snelling was thirty and his new bride fifteen. Snelling was a widower with a young son nearer to Abigail's age than his own.

14. Obst, "Abigail Snelling," 102, 106; Upham, *The Women and Children of Fort Saint Anthony*; Van Cleve, "Three Score Years and Ten," 27.

15. The 5th Regiment's sutler in 1819 was Mr. Devotion, who went broke in the early 1820s: Prescott, "Autobiography and Reminiscences," 35. Post Order no. 4, January 10, 1826, 5th Regiment Orderly Book, 1824-1828: "This council will ascertain the average number of men serving at this post in the months of September, Oct., Nov. and Dec. 1825 and determine what the tax on the Sutler shall be. They will determine whether any allowance of extra whiskey shall be made to the Bakers from the Post fund. They are requested to adopt some mode of selecting and processing the Books for the Post Library for which appropriations have been made. To form a code of regulations for the Librarian, to audit all accounts for the last year, which may be presented to them, and to act on all other subjects which are within the range of their duties, according to the Army regulations."

Account Books, 1825-1826, box 1, Bailly Papers; Obst, "Abigail Snelling," 104.

16. Account Books, 1825-1826, box 1, Bailly Papers.

17. The unstated implication of this line was that the entire Louisiana Purchase would eventually be colonized, despite the fact that much of the region was still home to Indigenous people in 1820.

18. Green, *A Peculiar Imbalance*, 5-8; Bachman, *The Last White House Slaves*, 10.

19. Bachman, *Northern Slave, Black Dakota*, 9, 12, 14-15. It is interesting to note that even if the military might defend the legality of the practice of slavery at the fort, the fur traders in Mendota were clearly violating both the Northwest Ordinance and the Missouri Compromise.

20. Green, *A Peculiar Imbalance*, 9. John Bliss recounts that following his father's assignment as commandant at the fort, the major's family stopped at St. Louis on their steamboat journey from Pennsylvania, where, "the last of our necessary purchases was made, to wit: a nice-looking yellow girl and an uncommonly black man": Bliss, "Reminiscences of Fort Snelling," 336. Captain John Garland, quartermaster at Fort Snelling in 1826, owned a young woman named Courtney whom he had brought with him from Fort Dearborn in Detroit: Bachman, *Northern Slave, Black Dakota*, 4-5, 14, 42.

21. Taliaferro, "Auto-biography," 254.

22. VanderVelde, *Mrs. Dred Scott*, ch. 3; Green, *A Peculiar Imbalance*, 10. Godfrey's story is told in Bachman, *Northern Slave, Black Dakota*.

23. Taliaferro journal, March 30, April 2, April 3, 1831; Bachman, *Northern Slave, Black Dakota*, 44.

24. Case, *The Relentless Business of Treaties*, 113-16.

25. Prucha, *The Sword of the Republic*, 199.

26. Nute, "Hudson's Bay Company Posts in the Minnesota Country," 283-87; Wingerd, *North Country*, 122-23.

27. Adams, "Early Days at Red River Settlement," 87.

28. Adams, "Early Days at Red River Settlement," 94, 111.

29. Martin, "Forgotten Pioneer," 19.

30. Wingerd, *North Country*, 124.

31. This and much of the following information regarding the families at Camp Coldwater is from a blog post by Bruce White: "Women of Coldwater: Native People at the Spring, MinnesotaHistory.net, January 25, 2010, http://www.minnesotahistory.net/wptest/?p=2159.

32. White, "Women of Coldwater"; Taliaferro journal, February 29, 1836.

33. Prescott, "Autobiography and Reminiscences," 36-37; Denial, *Making Marriage*, 79; Van Cleve, *"Three Score Years and Ten,"* 60-67. In trading directly with the Dakota the sutler was acting as a "factor," though Fort Snelling was not a part of the "factory system" of government trading posts, much to the chagrin of local fur traders, who complained to Taliaferro and Snelling about it: Alexis Bailly to Lawrence Taliaferro, February 16, 1831, Taliaferro Papers.

34. Woolworth, Dakota, Mixed Blood Indian, and White Biographical Files Notebook: Archibald John Campbell; Duncan Campbell; Atkins, *Creating Minnesota*, 28.

35. Atkins, *Creating Minnesota*, 29-30.

36. Document 19, Henry Atkinson to Lawrence Taliaferro, August 21, 1820; Document 24, Josiah Snelling to Lawrence Taliaferro, November 10, 1820—both Taliaferro Papers. Colin Campbell had worked as a Dakota interpreter for the British and in the fur trade on the St. Peter's River. He very likely had contacts with Sissetons at Big Stone Lake: Woolworth, Biographical Files.

37. Prescott, "Autobiography and Reminiscences," 34-35.

38. Document 25, Josiah Snelling to Lawrence Taliaferro, November 13, 1820, Taliaferro Papers.

39. Document 58, Josiah Snelling to Lawrence Taliaferro, October 19, 1824, Taliaferro Papers. It is not clear who "Markeemani" was. There was a Wahpeton leader named Mazamani, but as he had good relations with Taliaferro he was probably not responsible.

Note to Chapter 4 Sidebar, page 56

i. Green, *A Peculiar Imbalance*, 19-20; Wingerd, *North Country*, 237-38. Various historians have recorded the Dakota name of James Thompson's wife as Marpiyawicasta, but that name translates as "Cloud Man."

Notes to Chapter 5: Colonel Snelling's Fort

1. Heitman, *Historical Register and Dictionary of the United States Army*, 1:990, 653; Adams, "Early Days at Red River Settlement," 96.

2. Chernow, *Alexander Hamilton*, 117.

3. US Statutes at Large 2 (1789-1848), 359-72; 9th Congress, 1st Session, ch. 20, "An Act for Establishing Rules and Articles for the Government of the Armies of the United States," Article 25, 6.

4. Woodall, "William Joseph Snelling and the Early Northwest," 367-68.

5. Long, *The Northern Expeditions of Stephen H. Long*, 371.

6. Snelling journal, August 1825, Snelling Papers; Luecke and Luecke, *Snelling: Minnesota's First First Family*, 152-54.

7. Adams, "Early Days at Red River Settlement," 97.

8. Heitman, *Historical Register and Dictionary of the United States Army*, 557; Adjutant General Case File AA-34, NARG 153.

9. Prescott Reminiscences and Related Papers, 78-79.

10. Adjutant General Case I-39, 8-9, NARG 153.

11. Taliaferro journal, February 6, 1826; Prescott Reminiscences and Related Papers, 78-79. Prescott also notes that Baxley had a bad reputation and frequent conflicts regarding his wife, including killing the clerk of a Columbia Fur Company keelboat, whom the lieutenant believed was having an affair with her.

12. Adjutant General Case I-39, 7-16, NARG 153.

13. Taliaferro journals, January–March 1826; Order no. 4, January 5, 1826, 5th Regiment Orderly Book.

14. Adjutant General Case W-31, NARG 153. See also "A Soldier Disguised," in White, *The Tale of a Comet*, 1-22. The former lieutenant remained some months on the post as an assistant to the sutler, but died later in the year "while on an excursion at the mouth of the St. Croix, at the trading house of Mr. D. Campbell, on Saturday the 21st, Inst. [October 21, 1826]": October 27, 1826, 5th Regiment Orderly Book.

15. July 24, 1826, 5th Regiment Orderly Book.

16. US Articles of War (1806), Article 23, Section 148; Judge Advocate General Case I-39, 9, NARG 153.

17. Judge Advocate General Case U-54, 13-15, 17, NARG 153.

18. Reports of the Inspector General, Fort Snelling, August 1826, NARG 158, US War Department.

19. Reports of the Inspector General, Fort Snelling, August 1826, NARG 158, US War Department.

20. Taliaferro journal, April 23, 1826.

21. May 21, 1826, 5th Regiment Orderly Book.

22. Anderson, *Kinsmen of Another Kind*, 121.

23. Anderson, *Kinsmen of Another Kind*, 12-13, 121; Taliaferro journal, typescript, 32.

24. Document 58, Josiah Snelling to Lawrence Taliaferro, October 19, 1824, Taliaferro Papers.

25. Westerman and White, *Mni Sota Makoce*, 149, 151-52.

26. Westerman and White, *Mni Sota Makoce*, 149.

27. Taliaferro journal, September 19-20, 1825.

28. Taliaferro journal, January 6-7, April 30, 1826.

29. Taliaferro journal, January 13, 1826.

30. Subagent to Schoolcraft at "Michael's" (Madeline) Island; Taliaferro journal, May 7, May 26, 1827. Taliaferro felt he did a poor job of enforcing trade regulations: Taliaferro journal, January 18, 1828.

31. Westerman and White, *Mni Sota Makoce*, 115; Anderson, *Kinsmen of Another Kind*, 124.

32. Taliaferro journal, May 28, 1827; Snelling journal, May 30, 1827.

33. Special Order no. 94, May 29, 1827, 5th Regiment Orderly Book.

34. Anderson, *Kinsmen of Another Kind*, 125; Prescott Reminiscences and Related Papers, 87; Adams, "Early Days at Red River Settlement," 108-10.

35. Taliaferro journal, May 29-30, 1827.

36. Anderson, *Kinsmen of Another Kind*, 125; Prescott Reminiscences and Related Papers, 87.

37. Special Order no. 24, 1826, 5th Regiment Orderly Book.

38. Martin Zanger, "Red Bird," in Edmunds, ed., *American Indian Leaders*, 64-87; Taliaferro journal, July 9, 1827.

39. Orders no. 124, 125, July 9, 1827, 5th Regiment Orderly Book; Taliaferro journal, July 9, 1827.

40. Taliaferro journal, June 9, July 10, July 11, July 13, 1827. "Black Dog" was possibly not the individual's real name, as the chief of the Black Dog Village, near the fort, was frequently referred to as "Black Dog chief": Westerman and White, *Mni Sota Makoce*, 66.

41. September 10, 1827, Snelling Papers.

42. Luecke and Luecke, *Snelling: Minnesota's First First Family*, 194-95.

43. Judge Advocate General Case I-39, 7-16, NARG 153.

44. Judge Advocate General Case I-39, 38, NARG 153.

45. Judge Advocate General Case I-39, 2-3, NARG 153.

46. Consolidated Correspondence: Josiah Snelling, NARG 92, Office of the Quartermaster General, in Fort Snelling Papers, box 2, folder 17.

47. Hunter had a long army career. During the Civil War he was commander of the Department of the South and was among the first Union officers to call for the enlistment of Black soldiers: Hunter, *Report of the Military Services of Gen. David Hunter*, 7, 8.

48. Anderson, *Kinsmen of Another Kind*, 128. The understanding that the use of the land was a loan was made clear by Little Crow in several conversations with Taliaferro in 1829: see Westerman and White, *Mni Sota Makoce*, 144.

Notes to Chapter 5 Sidebars, pages 68, 71

i. Van Cleve, *"Three Score Years and Ten,"* 25.

ii. Josiah Snelling to Thomas Jesup, April 1, 1826, US War Department Quartermaster Generals Correspondence. Snelling had also complained of the blackbirds in a note to Commissary General of Subsistence George Gibson, May 16, 1823. It seems to have been a constant problem.

Notes to Chapter 6:
A Way of Life Unravels

1. Taliaferro journal, April 13, 1828.

2. Green, *A Peculiar Imbalance*, 9; Bachman, *The Last White House Slaves*, 16.

3. 21st Congress, 1st Session, House, Chapter 148, Section 6, May 28, 1830.

4. Prucha, *The Sword of the Republic*, 340.

5. Westerman and White, *Mni Sota Makoce*, 155.

6. Case, *The Relentless Business of Treaties*, 96.

7. Gilman, *Henry Hastings Sibley*, 35-37.

8. Gilman, *Henry Hastings Sibley*, 38.

9. Gilman, *Henry Hastings Sibley*, 34-36.

10. Wingerd, *North Country*, 96-100; Bachman, *Northern Slave, Black Dakota*, 18-19.

11. Anderson, *Kinsmen of Another Kind*, 139.

12. Taliaferro journal, September 8, 1830.

13. Gilman, "The Last Days of the Upper Mississippi Fur Trade," 123-40.

14. Taliaferro journal, May 20, June 17, 1829.

15. Westerman and White, *Mni Sota Makoce*, 104; Folwell, *A History of Minnesota*, 1:184-87.

16. Taliaferro journal, August 16, 1830; Westerman and White, *Mni Sota Makoce*, 106-7.

17. Taliaferro journal, May 28-31, 1832; Dr. John Emerson to Surgeon General Thomas Lawson, December 14, 1838, NARG 112, Office of the Surgeon General, selected letters received, Fort Snelling Papers, box 3.

18. Carroll, "Who Was Jane Lamont?" 189.

19. Carroll, "Who Was Jane Lamont?" 187, 192-93. It is interesting to note that much of what we know about Sibley's relationship with Helen comes from the papers of William Brown, in whose household she was raised. We know that Sibley and Helen corresponded, but it seems that Sibley's daughters by his American wife, Sarah Jane Steele, destroyed these letters before donating their father's papers to the Minnesota Historical Society.

20. Carroll, "Who Was Jane Lamont?" 189-90.

21. Bushnell, *Seth Eastman*, 23.

22. Wingerd, *North Country*, 145-50.

23. Wingerd, *North Country*, 111.

24. Taliaferro journal, July 7, 1834.

25. Anderson, *Kinsmen of Another Kind*, 168.

26. Westerman and White, *Mni Sota Makoce*, 105; Taliaferro journal, July 9, 1834.

27. McMahon and Karamanski, *North Woods River*, 76, 77-78.

28. Case, *The Relentless Business of Treaties*, 99.

29. Case, *The Relentless Business of Treaties*, 99.

30. Westerman and White, *Mni Sota Makoce*, 162-63.

31. Taliaferro journal, June 9, 1838.

32. Case, *The Relentless Business of Treaties*, 81; Letter of Acting Quartermaster General Trueman Cross to the Acting Secretary of War C. A. Harris, October 19, 1836, US War Department Quartermaster Generals Correspondence.

33. Prucha, "Army Sutlers and the American Fur Company," 26-27.

34. Prucha, "Army Sutlers and the American Fur Company," 28.

35. Taliaferro journal, January 29, 1836.

36. Samuel Stambaugh to Thomas Jesup, February 17, 1836; Samuel Stambaugh to Trueman Cross, September 7, 1836—both US War Department Quartermaster Generals Correspondence.

37. Letter of Acting Quartermaster General Trueman Cross to the Acting Secretary of War C. A. Harris, October 19, 1836; re: leases, December 1838, September 26, 1844, October 10, 1846—all US War Department Quartermaster Generals Correspondence.

38. Rorabaugh, *The Alcoholic Republic*, 9; Prucha, *The Sword of the Republic*, 324. In 1830 Americans consumed an

estimated equivalent of seven gallons of pure alcohol per person per year or ninety bottles of eighty-proof whiskey per year: Okrent, *Last Call*, 8.

39. Wingerd, *North Country*, 158.

40. Bliss, "Reminiscences of Fort Snelling," 337. Prisoners were usually kept in solitary confinement on bread and water: Bliss, "Reminiscences of Fort Snelling," 344.

41. Taliaferro journal, November 7, 1834, October 14, 1835.

42. White and White, *Fort Snelling in 1838*, 95. Depictions of the area from the 1830s show a region largely devoid of trees, far different from the late twentieth- and early twenty-first-century view. Standing at the Sibley House today, one cannot see Fort Snelling or Bdote through the heavy growth of timber. Strange as it may seem, this probably represents what an "undisturbed" landscape would have looked like.

43. Taliaferro journal, April 17, April 20, June 23, 1836. It's not clear what guns the agent was referring to. It seems unlikely the soldiers could loan out their army muskets without serious consequences. They more likely were loaning personal weapons, like shotguns used for hunting.

44. Taliaferro journal, March 14, 1831, July 19, 1834.

45. Taliaferro journal, September 7, 1830.

46. Wingerd, *North Country*, 158. Major Bliss was promoted to the rank of lieutenant colonel and transferred to the 6th Regiment. He and his family left the fort in May 1836.

47. Folwell, *A History of Minnesota*, 1:218; Jones, *Citadel in the Wilderness*, 200; US War Department Quartermaster Generals Correspondence, Brunson to Arnold Plummer, June 14, 1838. Captain Epharim Kirby Smith, not to be confused with his younger brother, Edmund Kirby Smith, also an army officer, who later became a prominent general in the Confederate Army. 40th Congress, 3rd Session, House Executive Documents, Epharim Kirby Smith to Joseph Plympton, October 19, 1837, serial 1372, 16.

48. Taliaferro journal, April 22, 1836.

49. Williams, *A History of the City of St. Paul*, 58–61.

50. Heitman, *Historical Register and Dictionary of the United States Army*, 225; Taliaferro journal, May 14, 1836.

51. Taliaferro journal, February 23, February 26, May 18, 1826, May 20, May 24, 1829, June 6, 1834, February 7, June 12, 1836; Adjutant General Correspondence, Lieutenant Colonel William Davenport to Adjutant General Roger Jones, May 27, 1827.

52. Taliaferro journal, July 9, 1838.

53. Loehr, "Franklin Steele, Frontier Businessman," 309–10. Steele's signature appears on a petition from the post dated July 21, 1837.

54. According to Taliaferro, Samuel Stambaugh seems to have made himself obnoxious: "This man of general notoriety happens to stand not only low at the Department [of Indian Affairs], but equally so with all honorable men": Taliaferro journal, October 14, 1838.

55. Folwell, *A History of Minnesota*, 1:452–54, relates the many variations on this story.

56. Loehr, "Franklin Steele, Frontier Businessman," 312.

57. Wingerd, *North Country*, 159; Prescott Reminiscences and Related Papers, 168; Taliaferro journal, September 21, July 21, 1838.

58. 27th Congress, 2nd Session, House Report 853, Serial 410, 3.

59. Westerman and White, *Mni Sota Makoce*, 127.

60. Wingerd, *North Country*, 160–61; Wingerd, *Claiming the City*, 19–20; Taliaferro journal, June 9, 1838.

61. Report to Governor Cass (?), July 4, 1838, Taliaferro Papers.

62. Taliaferro journal, June 3, 1836. The author is grateful to Tom Shaw for pointing out use of "My friends"; see Taliaferro journals, August 5, August 11, 1830, October 9, 1835.

63. Taliaferro, "Auto-biography," 227.

64. Taliaferro, "Auto-biography," 253–54.

Notes to Chapter 6 Sidebars, pages 87, 91, 99

i. Taliaferro journal, June 10, 1834; Wingerd, *North Country*, 91, 92; Goodman and Goodman, *Joseph R. Brown*, 58–59, 83, 99.

ii. Wingerd, *North Country*, 148; Taliaferro journal, July 1, July 3, 1831.

iii. Taliaferro journal, July 1831. Penichon and Black Dog Villages and Cloud Man's "Eatonville" were all located within a few miles of the fort: Westerman and White, *Mni Sota Makoce*, 126.

iv. White and White, *Fort Snelling in 1838*, 88, 91.

v. NARG 98, US Army Commands, 5th Regiment, box 4, Fort Snelling Papers; Prucha, *The Sword of the Republic*, 325, 326; US Army, Judge Advocate General, court-martial register and case files, August 21, 1834, MNHS microfilm M227.

vi. US Army, Judge Advocate General, court-martial register and case files, August 21, 1834, MNHS microfilm M227.

Notes to Chapter 7:
An Outpost and Source of Change

1. Wingerd, *North Country*, 104–6; Blegen, "The 'Fashionable Tour' on the Upper Mississippi," 378–79.

2. Taliaferro journal, October 10, 1835, June 2, 1839; Lehman, *Slavery's Reach*, 149–66.

3. White and White, *Fort Snelling in 1838*, 52.

4. Wingerd, *Claiming the City*, 20; Gilman, *Henry Hastings Sibley*, 77.

5. Gilman, *Henry Hastings Sibley*, 84–87.

6. Gilman, Gilman, and Miller, *The Red River Trails*, 2, 8; Folwell, *A History of Minnesota*, 1:236.

7. Gilman, *Henry Hasting Sibley*, 103–8.

8. At this time the ability to sketch maps and make drawings of terrain and fortifications were considered important skills for an officer.

9. Eastman, *Dahcotah*, vii.

10. Eastman, *Dahcotah*, iii, iv, x.

11. Eastman, *Dahcotah*, xvi.

12. Eastman, *Dahcotah*, x. "Manifest Destiny," a concept then current in the United States that history had a purpose or design revealed or "manifested" by particular events; the idea was commonly used to justify American expansion: Boyer, ed., *The Oxford Companion to United States History*, 470.

13. Bushnell, *Seth Eastman*, 25. Mary Eastman made reference to her husband having a daguerreotype "apparatus" that he used in his portrait work, but a decade would pass before a studio showcasing the new art form of photography appeared in St. Paul: Eastman, *Dahcotah*, xiv. This may have been the scene Mary Eastman described: Seth Eastman, *Dog Dance of the Dacotahs*, 1849, https://collections.artsmia.org/art/36165/dog-dance-of-the-dacotahs-seth-eastman.

14. Wingerd, *North Country*, 168; Post Returns, September 1844, NARG 94, US Adjutant General's Office, Returns from US Military Posts, 1800–1916, MNHS microfilm M195; unsigned journal of expedition from Fort Snelling beginning on September 18, 1844, NARG 98, US Army Commands, Selected Correspondence and Orders, Fort Snelling Papers, box 3.

15. Electus Backus to Henry Turner, February 28, 1845; Electus Backus to Henry Turner, April 1, 1845—both NARG 98, US Army Commands, Selected Correspondence and Orders, Fort Snelling Papers, box 3.

16. Major John B. Clark to A. A. A. Arthur, October 1, 1846, NARG 98, US Army Commands, Selected Correspondence and Orders, Fort Snelling Papers, box 4.

17. Post Returns, January, May, 1847; Seth Eastman to Henry Stanton, July 27, 1847, NARG 98, US Army Commands, Selected Correspondence and Orders, Fort Snelling Papers, box 3. Percussion cap muskets could be loaded more quickly and fired more reliably than flintlocks.

18. Wingerd, *North Country*, 220.

19. Stearns quoted in Prucha, "Fort Ripley," 206–9; Gustavus Otto to Gotlieb Hafner, August 18, 1849, Gustavus Otto letters.

Notes to Chapter 7 Sidebars, pages 110, 111, 112

i. Van Cleve, *"Three Score Years and Ten,"* 45.

ii. Goodman, *Paddlewheels on the Upper Mississippi*, 28.

iii. Beck, "Military Music at Fort Snelling," 78.

iv. Beck, "Military Music at Fort Snelling," 83.

v. Wingerd, *North Country*, 143; Westerman and White, *Mni Sota Makoce*, 164.

Notes to Chapter 8: Endings

1. *Minnesota Pioneer*, April 17, 1851, 2; *Minnesota Pioneer*, September 11, 1851, 2.

2. Beck, "Military Music at Fort Snelling," 229, 243–50.

3. Alexander Ramsey's initial attempt to meet with the Dakota in 1849 fell flat when Dakota leaders simply did not show up: Folwell, *A History of Minnesota*, 1:266, 273–74.

4. James Goodhue, dispatches from Traverse des Sioux, in Le Duc, *Minnesota Yearbook for 1852*, 47, as quoted in Wingerd, *North Country*, 194. For a detailed account of these treaties and their negotiation and an excellent discussion of the difficulties involved in translating the meaning of the treaties from English into Dakota, see Westerman and White, *Mni Sota Makoce*, 163–80.

5. *Minnesota Pioneer*, July 31, 1851, 2.

6. *Minnesota Pioneer*, May 1, 1851, 2; *Minnesota Pioneer*, June 19, 1851, 2. The "Turkish pantaloons" reference is to a women's rights movement to reform women's fashion away from corsets and skirts to more practical wear: see Lorraine Boissoneault, "Amelia Bloomer Didn't Mean to Start a Fashion Revolution, but Her Name Became Synonymous with Trousers," Smithsonian online, May 24, 2018, https://www.smithsonianmag.com/history/amelia-bloomer-didnt-mean-start-fashion-revolution-her-name-became-synonymous-trousers-180969164/. The new fashion was considered very daring.

7. See, for example, *Minnesota Democrat*, May 27, 1851.

8. Stephen Riggs to Seliah B. Treat, July 31, 1852, as quoted in Anderson, *Kinsmen of Another Kind*, 193; Wingerd, *North Country*, 196–204.

9. Hercules Dousman to F. B. Sibley, June 25, 1852, quoted in Wingerd, *North Country*, 203; *Minnesota Pioneer*, October 27, 1853, 2.

10. Upham, *Minnesota Geographic Names*, 2–3.

11. Williams, *A History of the City of St. Paul*, 358, 380. Bounty land warrants were essentially blank deeds for federal lands, much like scrip. They were first issued to veterans of the War of 1812 and then on a larger scale to veterans of the Mexican War. Depending on rank, a veteran received a given number of acres of federal lands that could be redeemed anywhere. Few actually settled or claimed their bounty lands; most warrants were sold, at a discount, to land speculators.

12. Kane, *The Waterfall that Built a City*, 36–37. On the map on page 127, "Rice Lake" is present-day Lake Hiawatha, and "Mother Lake" Lake Nokomis. The lakes identified as "Mud Lake" and "Duck Lake" no longer exist.

13. Ritchey, "Claim Associations and Pioneer Democracy in Early Minnesota," 93; Kane, *The Waterfall that Built a City*, 37. The 1841 Preemption Act allowed any male, single or head of household over age twenty-one, or widow who was also a US citizen or immigrant who had filed letters of intention to become a citizen, to make a "preemption" claim of federal lands, to which the "Indian land title at the time of such settlement had been extinguished" before any official public land sales. Those qualifying could purchase up to 160 acres of land at the standard rate of $1.25 per acre, providing they erected a "dwelling" on the property and made it their residence. This was a departure from previous federal land laws because it encouraged immigration rather than emphasizing land sales as a source of federal income: The Minnesota Legal History Project, "Pre-emption Act of 1841," http://www.minnesotalegalhistoryproject.org/archives.cfm?article=129.

14. Quartermaster Correspondence, Fort Snelling Papers, box 1, folder 1; *Minnesota Pioneer*, May 29, 1854.

15. Post returns from February through September 1855 indicate only 79 to 81 men available for duty in the garrison: NARG 94, US Adjutant General's Office, Returns from US Military Posts, 1800–1916, MNHS microfilm M195.

16. Gilman, "The Last Days of the Upper Mississippi Fur Trade," 136–40.

17. Post Returns, January 1857.

18. *Minnesota Pioneer*, November 24, 1853, 2.

19. White, "Indian Visits," 5–7.

20. Post Returns, January–July 1856.

21. Thomas Sherman to Samuel Medary, June 24, 1857, Territorial Governor Samuel Medary, Correspondence, Indian and Military Affairs, box 115.I.19.9b. Thomas Sherman was a career army officer and veteran of the Mexican War. He was not closely related to the more famous General William T. Sherman.

22. Larson, "A New Look at the Elusive Inkpaduta."

23. Beck, *Columns of Vengeance*, 6–8.

24. Beck, *Inkpaduta: Dakota Leader*, 54–55. By numerous accounts Inkpaduta continued to fight against the Americans for the next twenty years, with witnesses placing him at many major battles, including Killdeer Mountain and Little Bighorn: Larson, "A New Look at the Elusive Inkpaduta," 29–35.

25. 35th Congress, 1st Session, House Report 351, "Fort Snelling Investigation," 7.

26. "Fort Snelling Investigation," 8, 26.

27. "Fort Snelling Investigation," 16, 20.

28. *Chatfield Republican*, September 5, 1857.

29. "Fort Snelling Investigation," 20.

30. 35th Congress, 1st Session, House Miscellaneous Documents 133, "Report of a Board of Officers of the Army, as to the Necessity of Maintaining Fort Snelling as a Military Depot," 2, 3.

31. Folwell, *A History of Minnesota*, 1:509–13.

32. Captain A. W. Reynolds to Quartermaster General Thomas Jesup, June 29, 1858, quoted in Osman, *Fort Snelling and the Civil War*, 6.

Notes to Chapter 8 Sidebars, pages 120, 123, 128

i. All census information from Harpole and Nagle, eds., *Minnesota Territorial Census*, 11–17.

ii. Lewis A. Armisted was later a prominent Confederate general, mortally wounded at Gettysburg.

iii. Most of the infants, eight of them aged one or two years, seem to have been born at the fort.

iv. Bachman, *Northern Slave, Black Dakota*, 61.

v. *Minnesota Pioneer*, May 27, 1852, 2. It should be noted that this story was picked up and reprinted by newspapers both in the United States and in Great Britain.

vi. *Minnesota Pioneer*, April 28, 1853.

vii. *Minnesota Pioneer*, May 5, 1853.

Notes to Chapter 9: Conflict, Conflagration, and Tragedy

1. An Irishman by birth, James Shields immigrated in the 1830s to Illinois, where he practiced law and served in the legislature. He became a brigadier general of volunteers during the Mexican War. Following that conflict he acted briefly as governor of Oregon Territory in 1848, then as US senator from Illinois from 1849 to 1855. He was senator from Minnesota for less than a year: Folwell, *A History of Minnesota*, 2:7, 8.

2. Wingerd, *North Country*, 298; Anderson, *Kinsmen of Another Kind*, 247.

3. John B. Floyd to George L. Becker, November 11, 1858, War Department Correspondence; Folwell, *A History of Minnesota*, 2:53.

4. *Stillwater Messenger*, July 17, 1860.

5. *Chatfield Democrat*, October 6, 1860; *Stillwater Messenger*, October 6, 1860.

6. John Floyd was later accused of transferring weapons from arsenals and posts in northern states to arsenals located in the South, even as states were voting to leave the union. During the Civil War he served as a general in the Confederate Army: Swanberg, "Was the Secretary of War a Traitor?"

7. Quoted in Smith, "Minnesota on the Verge of Civil War," 59.

8. Osman, *Fort Snelling and the Civil War*, 9.

9. Osman, *Fort Snelling and the Civil War*, 217.

10. Moe, *The Last Full Measure*, 25–29. One of the cavalry regiments was referred to as "mounted rangers," and the two battalions—an organization of companies smaller than a regiment—were known by the names of their commanders, "Hatch's Battalion" and "Brackett's Battalion."

According to Dyers's *Compendium of the War of the Rebellion*, 11, the total number of troops serving in the Union Army from Minnesota was 24,020. The board of commissioners who compiled the rosters found in *Minnesota in the Civil and Indian Wars* included the "companies of citizen soldiers" temporarily formed as militia units to fight the Dakota, for a total of 26,717, a significant portion of the 1860 white population of 172,023: Minnesota Board of Commissioners, *Minnesota in the Civil and Indian Wars*, 1:vi.

11. Thomas Christie to James Christie, October 21, 1861, Christie Papers.

12. Osman, *Fort Snelling and the Civil War*, 29–31.

13. *Chatfield Democrat*, May 18, 1861.

14. Wills, *Boosters, Hustlers, and Speculators*, 100–109.

15. Wowinape's account was translated with the help of Stephen Riggs and published in 1891 by H. L. Gordon: see Anderson and Woolworth, eds., *Through Dakota Eyes*, 40–43.

16. Wingerd, *North Country*, 307. Regarding attacks on civilians, see: Carley, *The Sioux Uprising of 1862*, 21–24; Wingerd, *North Country*, 306–11; Lass, "Histories of the U.S.-Dakota War of 1862," 46–47, 53.

17. An accurate accounting of the number of settlers

killed is difficult to determine. The usual estimate is from five hundred to a thousand; Curtis Dahlin, who has made the most extensive study, in *Victims of the Dakota Uprising* estimates that six hundred were killed. Many people regarded as missing or killed may have fled the state and never returned; others might have died on the prairies and their bodies never found. In any event, the toll of dead was unprecedented.

18. Gilman, *Henry Hastings Sibley*, 173.

19. Cozzens, *General John Pope*, 16-17, 200-201.

20. *St. Paul Daily Press*, September 15, 1862, 1; Cozzens, *General John Pope*, 207.

21. Wingerd, *North Country*, 311; Gilman, *Henry Hastings Sibley*, 187. Sibley's forces had suffered reverses at the Battle of Birch Coulee, and at Acton on September 2 and 3, 1862. He had subsequently put off further movement until his men were better trained and equipped.

22. Bachman, *Northern Slave, Black Dakota*, 182, 185.

23. Wingerd, *North Country*, 321; Walt Bachman, "Colonel Miller's War," 110-12, in Bakeman and Richardson, eds., *Trails of Tears*; Elsie Cavender, "Army Brutality Marked Death March to Fort Snelling," *Granite Falls Tribune*, February 9, 1956, reprinted in Bakeman and Richardson, eds., *Trails of Tears*, 170-71. The following year Stephen Miller was elected governor of Minnesota.

24. Monjeau-Marz, *The Dakota Indian Internment at Fort Snelling*, 66, 70-71; Westerman and White, *Mni Sota Makoce*, 194; *St. Paul Daily Union*, November 22, 1863, 4. The woman gathering wood "some little distance from the encampment, was seized by a number of soldiers and brutally outraged. . . . It is a disgrace to the soldiers, a disgrace to the State, a disgrace to the country at large."

25. Cooke, "A Badger Boy in Blue," 99, quoted in Osman, *Fort Snelling and the Civil War*, 149. Regarding treatment of women, see Osman, *Fort Snelling and the Civil War*, 154.

26. Beck, *Columns of Vengeance*, 43-45; *St. Cloud Democrat*, November 13, 1862.

27. Alexander Ramsey, "Governor's Message," *Executive Documents of the State of Minnesota for 1862*, 12.

28. Monjeau-Marz, *The Dakota Indian Internment at Fort Snelling*, 100; Osman, *Fort Snelling and the Civil War*, 155.

29. Osman, *Fort Snelling and the Civil War*, 157, 159.

30. Millikan, "The Great Treasure of the Fort Snelling Prison Camp," 8; Osman, *Fort Snelling and the Civil War*, 151.

31. Millikan, "The Great Treasure of the Fort Snelling Prison Camp," 8-9.

32. Millikan, "The Great Treasure of the Fort Snelling Prison Camp," 9-11.

33. Millikan, "The Great Treasure of the Fort Snelling Prison Camp," 12, 15.

34. Canku and Simon, *The Dakota Prisoner of War Letters*, xix.

35. 37th Congress, 3rd Session, ch. 119, 1863, "An Act for the Removal of the Sisseton, Wahpaton, Medawakanton, and Wahpakoota Bands of Sioux or Dakota Indians, and for the Disposition of Their Lands in Minnesota and Dakota," https://www.loc.gov/law/help/statutes-at-large/37th-congress/session-3/c37s3ch119.pdf; *St. Paul Daily Press*, May 5, 1863.

36. Osman, *Fort Snelling and the Civil War*, 164-65.

37. Post Returns, September 1863.

38. Beck, *Columns of Vengeance*, 44-48, 51-52; *St. Paul Pioneer and Democrat*, November 8, 1862; Osman, *Fort Snelling and the Civil War*, 38.

39. Beck, *Columns of Vengeance*, 64-65, 70.

40. Folwell, *A History of Minnesota*, 2:276-77.

41. Beck, *Columns of Vengeance*, 154-57. Samuel J. Brown, an interpreter with the expedition, described the attack as "a perfect massacre . . . he pitched into their camp and just slaughtered them": Folwell, *A History of Minnesota*, 2:280.

42. Anderson, *Little Crow: Spokesman for the Sioux*, 7-8; Beck, *Columns of Vengeance*, 171.

43. Osman, *Fort Snelling and the Civil War*, 231-33; Folwell, *A History of Minnesota*, 2:296-97. Alfred Sully claimed to have killed over 150 warriors; the Lakota account says 31 were killed: Clodfelter, *The Dakota War*, 176.

44. Folwell, *A History of Minnesota*, 2:290.

45. *St. Paul Pioneer Press* quoted in Folwell, *A History of Minnesota*, 2:450; Osman, *Fort Snelling and the Civil War*, 172-73. There is an extensive description of the abduction and trial in Folwell, *A History of Minnesota*, appendix 16, 2:443-50.

46. Taylor, *African Americans in Minnesota*, 7.

47. *Stillwater Messenger*, May 12, 1863, 2, col. 2.

48. *St. Paul Press*, May 6, 1863. 2.

49. *St. Paul Pioneer*, May 6, 1863, 4. These papers representing rival parties carried on a similar line of vitriolic, racial, and ethnic attacks in 1858. Oddly enough, they merged a decade later.

50. *St. Paul Daily Press*, May 15, 1863, 4, col. 1. The Indians referred to were probably the Ho-Chunk.

51. Green, *A Peculiar Imbalance*, 138-39. Robert Hickman was a remarkable man. Literate, a fine orator, and a licensed preacher, he is said to have led a mass escape of slaves from Boone County, Missouri, by raft down the Missouri River. They had just entered the Mississippi when the *Northerner*, headed for St. Paul, picked them up and took them in tow. Rev. Hickman was a leading force in St. Paul's Black community in the nineteenth century. His followers referred to themselves as "Pilgrims" and founded Pilgrim Baptist Church in St. Paul, the first African American congregation in the state.

52. Osman, *Fort Snelling and the Civil War*, 194-200.

53. Osman, *Fort Snelling and the Civil War*, 244-46. This was the first compulsory service in US history. One could avoid service by paying a $300 commutation fee or by hiring a substitute.

54. Post Returns, July 1864.

55. Post Returns, March 1866.

56. Hyman, "Survival at Crow Creek."

57. Westerman and White, *Mni Sota Makoce*, 202-3.

Notes to Chapter 9 Sidebars, pages 141, 151

i. Thomas Saunders to M. C. Meigs, September 14, 1861, US War Department Quartermasters Generals Correspondence.

ii. DeCarlo, *Fort Snelling at Bdote*, 54; Osman, *Fort Snelling and the Civil War*, 160.

iii. "Concentration Camps, 1933–39," Holocaust Encyclopedia, US Holocaust Memorial Museum, https://encyclopedia.ushmm.org/content/en/article/concentration-camps-1933-39; Andrea Pitzer, "Concentration Camps Existed Long Before Auschwitz," Smithsonian online, November 2, 2017, https://www.smithsonianmag.com/history/concentration-camps-existed-long-before-Auschwitz-180967049/; Paul J. Springer, "The Term 'Concentration Camp' in Historical Perspective," Foreign Policy Research Institute, June 27, 2019, https://www.fpri.org/article/2019/06/the-term-concentration-camp-in-historical-perspective/.

iv. Osman, *Fort Snelling and the Civil War*, 160–61.

Notes to Chapter 10:
Expansion and a New Role

1. Montgomery Meigs to Edwin Stanton, June 16, 1866, US War Department Quartermaster Generals Correspondence.

2. Governor William Marshall to Secretary of War Edwin Stanton, July 23, 1866, US War Department Quartermaster Generals Correspondence.

3. Prosser, *Rails to the North Star*, 12, 17.

4. Samuel Holobird to Daniel Rucker, April 16, 1868, US War Department Quartermaster Generals Correspondence.

5. Folwell, *A History of Minnesota*, 1:514–15.

6. Robinson, *A Good Year to Die*, 16, 17, 24, 25.

7. Robinson, *A Good Year to Die*, xxv, xxvi.

8. Robinson, *A Good Year to Die*, 21–23.

9. Robinson, *A Good Year to Die*, 29–30.

10. Some Santee or Eastern Dakota, including Iŋkpaduta, were said to have been there as well: Viola, *Little Bighorn Remembered*, 4; Larson, "A New Look at the Elusive Inkpaduta," 24.

11. Robinson, *A Good Year to Die*, 216.

12. 45th Congress, 3rd Session, Senate Executive Document 36, vol. 1828.

13. *St. Paul Daily Globe*, October 19, 1880, 4, col. 5. Bridal Veil, a small but picturesque waterfall, is on the St. Paul side of the Mississippi between the current Marshall Avenue/Lake Street and Franklin Avenue bridges. Now mostly obscured, it is visible from the river trails and was a popular picnic spot in the later nineteenth century.

14. 46th Congress, 2nd Session, Senate Executive Document 54.

15. 45th Congress, 2nd Session, Senate Executive Document 73.

16. *Minneapolis Tribune*, April 5, 1878, 4, col. 3; *St. Paul Daily Globe*, March 23, 1878, 1.

17. *St. Paul Daily Globe*, December 11, 1878, 1; March 27, 1880, 1.

18. Summit Envirosolutions and Two Pines Resource Group, "The Cultural Meaning of Coldwater Spring," 20.

19. DeCarlo, *Fort Snelling*, 63.

20. Nankivell, *The History of the Twenty-Fifth Regiment*, 11–12.

21. *St. Paul Daily Globe*, May 31, 1883, 2, col. 5.

22. Wingerd, *Claiming the City*, 77; Green, "Minnesota's Long Road to Black Suffrage," 83–84; Green, *A Peculiar Imbalance*, 169–72; *St. Paul Daily Globe*, June 5, 1883, 2, col 4.

23. *St. Paul Daily Globe*, May 18, 1879. Both the *Globe* and the *Tribune* ran weekly columns on military matters, frequently reprinting general orders of the department and specific items relating to Fort Snelling.

24. 50th Congress, 1st Session, House Executive Document 344; *St. Paul Globe*, October 4, 1903.

25. *Chatfield Democrat*, July 17, 1891, 3, col. 4.

26. *Minneapolis Tribune*, April 30, 1894, 1, col. 5.

27. *Harpers Weekly*, May 11, 1895, in MHS Scrapbooks, vol. 12, 57–58.

28. Holbrook, *Minnesota in the Spanish-American War*, 107.

29. Holbrook, *Minnesota in the Spanish-American War*, 110.

30. Matsen, "The Battle of Sugar Point," 270.

31. The petition is reprinted in Roddis, "The Last Indian Uprising in the United States," 277.

32. Roddis, "The Last Indian Uprising in the United States," 272–73.

33. 53rd Congress, 3rd session, House Executive Document 174, Report of 3rd Regiment Surgeon W. C. Borden, 11.

34. 53rd Congress, 3rd session, House Executive Document 174, Army Medical Inspector report, 13.

35. 53rd Congress, 2nd Session, House Report 753; *St. Paul Pioneer Press*, August 2, 1903; MHS Scrapbooks, vol. 334, 46.

36. "Bridge Bill Is in Good Hands," *St. Paul Pioneer Press*, February 24, 1906.

37. *St. Paul Globe*, October 9, 1904; *St. Paul Dispatch*, November 1, 1904; *Minneapolis Journal*, October 29, 1904; regarding Congressman Frederick Stevens, *St. Paul Dispatch*, April 19, 1905.

38. *St. Paul Pioneer Press*, September 12, 1909; *Minneapolis Journal*, September 23, 1905.

39. *St. Paul Globe*, March 6, 1903.

40. "History of the United States Army," Twentieth Century, Wikipedia, https://en.wikipedia.org/wiki/History_of_the_United_States_Army#Twentieth_century.

41. *Finance and Commerce of the Twin Cities*, October 31, 1912, 1.

42. *Minneapolis Tribune*, January 1, 1912, 5.

Note to Chapter 10 Sidebar, page 167

i. 45th Congress, 3rd Session, Senate Report 770.

Notes to Chapter 11: World War

1. Holbrook and Appel, *Minnesota in the War With Germany*, 1:308–9.

2. Holbrook and Appel, *Minnesota in the War With Germany*, 1:181.

3. Holbrook and Appel, *Minnesota in the War With Germany*, 1:186.

4. Carol R. Byerly, "War Losses (USA)," 1914–1918-Online, International Encyclopedia of the First World War, https://encyclopedia.1914-1918-online.net/article/war_losses_usa.

5. 1914–1918-Online, International Encyclopedia of the First World War, https://encyclopedia.1914-1918-online.net/article/war_losses_usa. Most of these hospital admissions were for disease or accident.

6. *Duluth Herald*, October 9, 1918, 13; *Minneapolis Morning Tribune*, November 11, 1918, 7.

7. Fort Snelling General Hospital, *Reveille: The Call to a New Life* 29, no. 1, May 8, 1919, 4–5.

8. *Reveille*, May 8, 1919, 4–5.

9. "Women Nurses in the Civil War," *Understanding War Through Imagery: The Civil War in American Memory*, US Army Heritage and Education Center, https://ahec.armywarcollege.edu/exhibits/CivilWarImagery/Civil_War_Nurses.cfm.

10. Vane and Marble, "Contributions of the U.S. Army Nurse Corps in World War I."

11. Vane and Marble, "Contributions of the U.S. Army Nurse Corps in World War I"; *Reveille* 7, June 5, 1919, 14–15.

12. *Reveille*, May 8, 1919, 7.

13. *Reveille*, May 29, 1919, 5–6.

14. *Reveille*, June 12, 1919, 14–15; *Minneapolis Tribune*, March 23, 1919, 13.

15. *Reveille*, April 24, 1919, 7, 8.

16. *Willmar Tribune*, January 15, 1919, 1, 3.

17. *Little Falls Herald*, March 21, 1919, 3.

18. *Minneapolis Morning Tribune*, June 30, 1919, 6.

19. US Department of Veterans Affairs, "Minneapolis VA Health Care System," https://www.minneapolis.va.gov/about/history.asp.

20. US Army Medical Department, "Statistical Data, United States Army General Hospital No. 29, Fort Snelling, Minn.," 580, https://history.amedd.army.mil/booksdocs/wwi/MilitaryHospitalsintheUS/chapter27page580.pdf; Osman, *Fort Snelling Then and Now*, 3–4.

21. *Minneapolis Tribune*, March 23, 1919, 13.

Notes to Chapter 12: The Country Club of the Army

1. *Minneapolis Morning Tribune*, August 15, 1920, 12; September 2, 1921, 12. This was undoubtedly Dietrich Lange, who wrote *On the Trail of the Sioux*, a fictional work set during the Dakota War featuring two boy scouts. Lange was a supporter of the Boy Scout movement. See Minnesota Author Biographies, https://collections.mnhs.org/mnauthors/10001318.

2. *Minneapolis Morning Tribune*, September 12, 1920, 1.

3. *Minneapolis Morning Tribune*, September 5, 1920, State Fair Section, 3, 6.

4. Stewart, ed., *American Military History*, 2:54.

5. Weigley, *History of the United States Army*, 275–81.

6. Weigley, *History of the United States Army*, 321–23; *Irish Standard*, February 3, 1912, 1.

7. Weigley, *History of the United States Army*, 396–400.

8. *Minneapolis Morning Tribune*, October 4, 1921, 6, col. 1; *Fort Snelling Bulletin*, November 20, 1931, 1, 5.

9. *Fort Snelling Bulletin*, November 13, 1931, 2; May 29, 1937, 2. The writer was W. C. Nichols, vice president of the Northwestern Milling Company. Some of these men died at the post hospital or the Veterans Hospital at the fort: see *Fort Snelling Bulletin*, December 13, 1929, 1; October 8, 1933.

10. *Fort Snelling Bulletin*, January 25, 1929. 1; September 6, 1929, 1.

11. *Fort Snelling Bulletin*, April 25, 1930, 2.

12. *Fort Snelling Bulletin*, September 13, September 27, 1929; September 5, 1930, 2; March 8, 1935, 1.

13. *Fort Snelling Bulletin*, December 7, 1928, 1; January 4, 1929, 3; October 11, 1929, 3; March 20, 1931.

14. US Air Force, "General Walter C. Sweeney Jr.," https://www.af.mil/About-Us/Biographies/Display/Article/105473/general-walter-c-sweeney-jr/. Sweeney later became an important figure in the service during the interwar years, rising to the rank of major general in command of the 3rd Division in 1938.

15. *Fort Snelling Bulletin*, May 31, 1929, 1.

16. *Fort Snelling Bulletin*, June 27, 1930, 2; "United States Army's Family and MWR Programs," Wikipedia, https://en.m.wikipedia.org/wiki/United_States_Army%27s_Family_and_MWR_Programs.

17. *Fort Snelling Bulletin* 1, no. 4 (n.d.): 1; April 29, 1932, 1; *Minneapolis Morning Tribune*, July 8, 1929, 10.

Built in the early twentieth century as a place for cavalry and artillery to drill during inclement weather, the Riding Hall, a large structure with copious amounts of open space, was used for many purposes during Fort Snelling's days as an army post. It still stands and is actively used by the Boy Scouts.

The reference to Fort Snelling as "the country club of the army" seems to have come from local newspapers, but its original source is unclear. A "Historical Resume of the Fort Snelling Officers' Country Club" is microfilmed with the March 30, 1934, issue of the *Fort Snelling Bulletin* and also printed on page 3. Apparently written for the club's dedication on the following day, it gives an account of the building of the club in 1933–34, concluding, "It is hoped and believed that this fine building will serve as a real Country Club and be a social center and place of assembly and friendly contact for all Officers and Ladies, who may be stationed here, and their friends in St. Paul and Minneapolis, which cities have so long shown the greatest friendship and interest in the welfare of Fort Snelling."

18. *Fort Snelling Bulletin*, January 17, 1930, 1; US Army

Quartermaster Corps, Fort Snelling Building Records, vol. 1.

19. Olson, *Electric Railways of Minnesota*, 54. In 1909 a line was completed from West Seventh Street in St. Paul across an improved Fort Snelling bridge. It passed near the round tower and proceeded north to Minneapolis. Autos, their care and storage, received occasional mention in the *Fort Snelling Bulletin*, as, sadly, did fatal accidents involving soldiers: see *Fort Snelling Bulletin*, January 4, 1929, 1.

20. *Minneapolis Tribune*, October 8, 1928; *Fort Snelling Bulletin*, June 7, 1929, 1; April 3, 1937, 1.

21. *Minneapolis Journal*, September 15, October 8, 1928.

22. *Fort Snelling Bulletin*, September 16, 1932, 1.

23. *Minneapolis Tribune*, April 16, 1933, 27.

24. Stewart, ed., *American Military History*, 2:65.

25. Stewart, ed., *American Military History*, 2:65.

26. *Fort Snelling Bulletin*, May 19, 1933, 1.

27. Cameron, "Civilian Conservation Corps in Minnesota."

28. *Fort Snelling Bulletin*, April 28, 1933, 1.

29. *Fort Snelling Bulletin*, June 16, 1933, 4; January 14, 1935, 3; April 27, 1935, 1.

30. *Fort Snelling Bulletin*, October 6, 1936, 1; *Minneapolis Tribune*, October 18, 1935.

31. *Minneapolis Tribune*, February 5, 1926, 18; *Fort Snelling Bulletin*, July 8, 1939, 1.

32. *Fort Snelling Bulletin*, May 23, 1936, 1; August 22, 1936, 1, 4.

33. *Fort Snelling Bulletin*, August 9, 1929, 1.

34. Weigley, *History of the United States Army*, 342-43, 396-400.

35. Osman, *Fort Snelling Then and Now*, 4.

Notes to Chapter 12 Sidebars, pages 207, 208

i. *Fort Snelling Bulletin*, February 23, 1934, 1.
ii. Slovak, "Smartest Horse in the U. S. Army."

Notes to Chapter 13:
World War II

1. Weigley, *History of the United States Army*, 426-28.

2. *Minneapolis Tribune*, August 24, 1940, 1. The same front page had a dramatic story of the ongoing Battle of Britain.
Minneapolis Star-Journal, November 3, 1940, 1; Osman, "Fort Snelling's Last War," in *Fort Snelling Then and Now*, 3. Army war plans for the defense of North America, known as the Rainbow plans, had begun in 1934 and Rainbow I was finalized in September 1939. It seems logical to assume that Fort Snelling's designation as a recruitment and induction center had already been determined: Stewart, ed., *American Military History*, 2:74.

3. Osman, "Fort Snelling's Last War," 3.

4. *Fort Snelling Bulletin*, March 21, 1941, 1; *Minneapolis Star Journal*, April 12, 1941, 9.

5. Stewart, ed., *American Military History*, 2:75; Churchill, *The Grand Alliance*, 606.

6. *Fort Snelling Bulletin*, December 12, 1941, 1; *Minneapolis Tribune*, December 8, 1941, 1.

7. Military Occupational Classification of Enlisted Personnel, War Department Technical Manual TM 12-427 (Washington: GPO, 1944); Osman, *Fort Snelling Then and Now*, 9-10.

8. Osman, *Fort Snelling Then and Now*, 10; *Fort Snelling Bulletin*, June 17, 1946, 4.

9. "3rd U.S. Infantry Regiment (The Old Guard)," Interwar Period (1919-39), Wikipedia, https://en.wikipedia.org/wiki/3rd_U.S._Infantry_Regiment_(The_Old_Guard)#Interwar_period_(1919-39). For the remaining years of World War II, the resident troops were the military police (MPs) and those staffing the War Department Personnel Center: *Fort Snelling Bulletin* 21 (1946): 2.

10. *Fort Snelling Bulletin*, January 9, 1942, 1. The United Service Organizations is a nonprofit organization founded in 1941 to bring entertainment and support to American service members at home and overseas.

11. *Fort Snelling Bulletin*, February 27, 1942, 1; March 6, 1942, 10.

12. *Fort Snelling Bulletin*, December 15, 1945, 1. Bing Crosby seemed to be everyone's favorite singer. For an example, the "Believe It or Not" cartoon, a satire of a well-known newspaper feature, in the April 16, 1943, issue featured a clever version of a military zoot suit.

13. Osman, *Fort Snelling Then and Now*, 12.

14. Osman, *Fort Snelling Then and Now*, 12; *Minneapolis Star Journal*, October 6, 1942, 12.

15. Ano, "Loyal Linguists," 274, 275.

16. Ano, "Loyal Linguists," 278.

17. *Fort Snelling Bulletin*, April 26, 1946, 1.

18. *Fort Snelling Bulletin*, June 21, 1946, 4; *Minneapolis Morning Tribune*, May 10, 1945, 6.

19. *Minneapolis Morning Tribune*, December 28, 1945, 1; *Fort Snelling Bulletin*, June 21, 1946, 1.

20. *Minneapolis Star*, October 15, 1946, 17.

Note to Chapter 13 Sidebar, page 219

i. Chuck Rowland is quoted in Van Cleve, *Land of 10,000 Loves*, 46; the original interview with Allan Bérebé, March 19, 1984, is at the GLBT Historical Society, San Francisco. Van Cleve notes that he excerpted the comments from a transcript by David Hughes in 2009 and edited them for clarity.

Notes to Chapter 14:
The Making of Historic Fort Snelling

1. *Minneapolis Journal*, June 29, 1946; *Minneapolis Morning Tribune*, September 7, 1946, 6.

2. *Minneapolis Morning Tribune*, January 26, 1946, 1; Expansion files, 1945-1946, Wold-Chamberlain Field files, Metropolitan Airports Commission.

3. "A Brief Report on Aviation Progress and Airport Problems in the Twin Cities, by the Minneapolis-St. Paul Airports Commission," January 30, 1957, 7–9, MNHS; Harper, "The Minneapolis–St. Paul Metropolitan Airports Commission," 366, 437; Expansion files, 1947–1950, revised expansion plan, Wold-Chamberlain Field Expansion Plan, October 13, 1949, file 1954–1955 MAC Resolution 293, Wold-Chamberlain Field files, Metropolitan Airports Commission. For a more personal view of this era, see Watson, "Fort Snelling as I Knew It."

4. *Minneapolis Star*, April 25, 1951, 1; *Minneapolis Morning Tribune*, May 1, 1951, 1. The air force did receive thirty-five acres of the Fort Snelling property next to the runway for use by the air national guard.

5. US Army Quartermaster Corps, Fort Snelling Building Records, box 1, vol. 1.

6. *Minneapolis Morning Tribune*, November 27, 1939; May 24, 1940; Johnson, "Reconstructing Old Fort Snelling," 84.

7. Fridley, "Fort Snelling," 190.

8. Fridley interview, 112–13.

9. Fridley, "Fort Snelling," 191.

10. Fridley interview, 114.

11. MNHS Director's Annual Report, 1957, MNHS Archives.

12. Russell W. Fridley, "Fort Snelling: The Beginning of Minnesota," n.d. (circa 1958), MNHS Archives, box 1056. This may have been the director's annual meeting address.

13. Fridley interview, 85, 86.

14. Fridley interview, 89: "One of my rationales for taking the Society into this area is it seemed to me that such a program was not only needed, but that it would give the Society a statewide base." Fridley's considerable personal skills undoubtedly facilitated this process. He was a natural politician. In his interview he in turn gave credit to staff members June Holmquist and Jean Brookins for their work on *Minnesota's Major Historic Sites: A Guide* (St. Paul: MNHS, 1963).

15. Fridley interview, 86–87.

16. Johnson, "Reconstructing Old Fort Snelling," 83.

17. "Archaeology at Fort Snelling," Historic Fort Snelling, https://www.mnhs.org/fortsnelling/learn/archaeology.

18. Johnson, "Reconstructing Old Fort Snelling," 86, 88.

19. Johnson, "Reconstructing Old Fort Snelling," 86, 88, 89.

20. Stephen Osman, unpublished overview, "Living History Evolution and Change at a Historic Site," June 1, 1989, Historic Fort Snelling staff files.

21. Osman, "Living History," 3.

22. Osman, "Living History," 4.

23. Nancy Eubank, Fort Snelling Visitor's Center Interpretation—Planning Document, October 21, 1980, 2, MNHS Archives, box 147.E.19.3B.

24. Eubank, Fort Snelling Visitor's Center Interpretation—Planning Document, 8–14. This information was provided to the author by current permanent Fort Snelling staff members.

25. Tyson, *The Wages of History*, 4.

26. White and White, *Fort Snelling in 1838*.

27. Historic Fort Snelling training documents, "Soldier Life Interpretive Station Treatment," February 1, 2019, 5.

28. "Waziyatawin: Take Down Fort Snelling," Minnesota Public Radio, June 28, 2016, https://www.mprnews.org/story/2016/06/28/books-what-does-justice-look-like.

29. For much of its history, the Minnesota Historical Society reflected the views of the white settler community that founded it and was largely insensitive to the perspective and feelings of the Indigenous people here. For example, the remains of Little Crow were on public display for many years and only returned to his family in the 1980s. In recent decades the society has worked with Indigenous people to create a better relationship based on trust and mutual respect.

30. Bakeman, "It's Time for Truth Telling at Fort Snelling"; Osman, "Fort Is Sacred to the Memory of Those Who Served There," 8.

31. For a more detailed discussion, see Bruce White, "Tearing Down Fort Snelling—Why It Makes Sense," MinnesotaHistory.net, April 11, 2009, http://www.minnesotahistory.net/wptest/?p=1339.

Notes to Chapter 14 Sidebars, pages 226, 230

i. Lovelace, *Early Candlelight*, xiv–xv, xvii; *Fort Snelling Bulletin*, September 27, 1949, 1. Though *Early Candlelight* was a success for Maud Hart Lovelace, she gained still greater fame for her series of children's and young adult books based on her childhood in Mankato, Minnesota, the Betsy-Tacy novels. They remain popular with young readers.

ii. Meyer, *Everyone's Country Estate*, 200–205.

Bibliography

Primary Sources

MANUSCRIPT COLLECTIONS

Alexis Bailly Papers. MNHS.

Jacob J. Brown Papers. Library of Congress. Available at MNHS.

William R. Brown Papers. MNHS.

Thomas and Carmelite Christie and Family Papers. MNHS.

Department of Transportation. Airport Commission Files. Minnesota State Archives.

Fort Snelling Papers. P333. Microfilm M715. MNHS.

Russell W. Fridley. Interview by Lucile M. Kane and Rhoda R. Gilman. OH82.6. MNHS.

Historic Sites Files. Minnesota State Archives.

Records of Territorial Governor Samuel Medary. MNHS.

MHS Scrapbooks, 1861-1922. MNHS.

Gustavus Otto letters. MNHS.

Philander Prescott Reminiscences and Related Papers. MNHS.

Henry Rowe Schoolcraft Papers. Library of Congress. Available at MNHS.

Henry H. Sibley Papers. MNHS.

Josiah Snelling Papers. MNHS.

Lawrence Taliaferro Papers. MNHS.

US Adjutant General's Office. Letters Sent by the Office of the Adjutant General. NARG 94. Microfilm M196. MNHS.

US Adjutant General's Office. Registers of Enlistments in the US Army, 1798-1897. NARG 98. Microfilm M191. MNHS.

US Adjutant General's Office. Returns from US Military Posts, 1800-1916. NARG 94. Microfilm M195. MNHS.

US Adjutant General's Office. Surgeons Quarterly Reports of Sick and Wounded, Fort Snelling. NARG 94. Microfilm M224. MNHS.

US Army, 5th Regiment Infantry. Order Books, 1819-1828. MNHS.

US Army, Office of the Judge Advocate General. Court-martial registers and case files (selected). NARG 153. Microfilm M227. MNHS.

US Congress. House and Senate Executive Documents.

US War Department. Letters Received by the Secretary of War, Main Series. NARG 107. Microfilm M243. MNHS.

US War Department. Quartermaster Generals Correspondence. Fort Snelling 1819-1868. NARG 92. Microfilm M222. MNHS.

US War Department. Secretary's Office. Letters Sent, Military Affairs. NARG 107. Microfilm M152. MNHS.

Woolworth, Alan R. Dakota, Mixed Blood Indian, and White Biographical Files Notebook. 1984. MNHS.

NEWSPAPERS

Chatfield Democrat
Chatfield Republican
Fort Snelling Bulletin
Granite Falls Tribune
Irish Standard
Little Falls Herald
Minneapolis Journal
Minneapolis Star
Minneapolis Star Journal
Minneapolis Tribune
Minnesota Democrat
Minnesota Pioneer
St. Cloud Democrat
St. Paul Dispatch
St. Paul Globe
St. Paul Pioneer
St. Paul Pioneer and Democrat
St. Paul Pioneer Press
St. Paul Press
St. Paul Union
Stillwater Messenger
Willmar Tribune

Secondary Sources

Adams, Ann. "Early Days at Red River Settlement, and Fort Snelling: Reminiscences of Ann Adams, 1821-1829." *Collections of the Minnesota Historical* 6 (1894).

Anderson, Gary Clayton. *Kinsmen of Another Kind: Dakota-White Relations in the Upper Mississippi Valley, 1650-1862.* 1984. Reprint, St. Paul: MNHS Press, 1997.

———. *Little Crow: Spokesman for the Sioux.* St. Paul: MNHS Press, 1986.

Anderson, Gary Clayton, and Alan R. Woolworth, eds. *Through Dakota Eyes: Narrative Accounts of the Minnesota Indian War of 1862.* St. Paul: MNHS Press, 1988.

Ano, Masaharu. "Loyal Linguists: Nisei of World War II Learned Japanese in Minnesota." *Minnesota History* 45, no. 7 (Fall 1977): 273-87.

Atkins, Annette. *Creating Minnesota: A History from the Inside Out.* St. Paul: MNHS Press, 2007.

Bachman, Walt. *The Last White House Slaves: The Story of Jane, President Zachary Taylor's Enslaved Concubine.* N.p.: The Author, 2019.

———. *Northern Slave, Black Dakota: The Life and Times of Joseph Godfrey.* Bloomington, MN: Pond Dakota Press, 2013.

Bakeman, Mary. "It's Time for Truth Telling at Fort Snelling." MinnPost, May 1, 2017. https://www.minnpost.com/community-voices/2017/05/it-s-time-truth-telling-fort-snelling/.

Bakeman, Mary H., and Antona H. Richardson, eds. *Trails of Tears: Minnesota's Dakota Indian Exile Begins.* Roseville, MN: Prairie Echoes, 2008.

Beck, Paul N. *Columns of Vengeance: Soldiers, Sioux, and the Punitive Expeditions, 1863-1864.* Norman: University of Oklahoma Press, 2013.

———. *Inkpaduta: Dakota Leader.* Norman: University of Oklahoma Press, 2008.

Beck, Roger Lawrence. "Military Music at Fort Snelling, Minnesota, from 1819 to 1858: An Archival Study." PhD diss., University of Minnesota, 1987.

Blegen, Theodore C. "The 'Fashionable Tour' on the Upper Mississippi." *Minnesota History* 20, no. 4 (Winter 1939): 377-96.

Bliss, John H. "Reminiscences of Fort Snelling." *Collections of the Minnesota Historical Society* 6 (1894).

Boyer, Paul S., ed. *The Oxford Companion to United States History.* Oxford: Oxford University Press, 2006.

Bray, Edmund C. "Millions of Years in the Making: The Geological Forces that Shaped St. Paul." *Ramsey County History* 32, no. 2 (Summer 1997): 13-15.

Brown, William L. III. *The Army Called It Home: Military Interiors of the Nineteenth Century.* Gettysburg, PA: Thomas Publications, 1992.

Bushnell, David I. *Seth Eastman: The Master Painter of the North American Indian.* Washington, DC: Smithsonian, 1932.

Cameron, Linda A. "Civilian Conservation Corps in Minnesota, 1933-1942." MNopedia. https://www.mnopedia.org/civilian-conservation-corps-minnesota-1933-1942.

Canku, Clifford, and Michael Simon. *The Dakota Prisoner of War Letters: Dakota Kaškapi Okicize Wowapi.* St. Paul: MNHS Press, 2013.

Carley, Kenneth. *The Sioux Uprising of 1862.* St. Paul: MNHS Press, 1979.

Carroll, Jane Lamm. "Who Was Jane Lamont? Anglo-Dakota Daughters in Early Minnesota." *Minnesota History* 59, no. 5 (Spring 2005): 184-96.

Case, Martin. *The Relentless Business of Treaties: How Indigenous Land Became US Property.* St. Paul: MNHS Press, 2018.

Chernow, Ron. *Alexander Hamilton.* New York: Penguin, 2004.

Churchill, Winston. *The Grand Alliance.* Boston: Houghton Mifflin, 1950.

Clodfelter, Micheal. *The Dakota War: The United States Army Versus the Sioux, 1862-1865.* Jefferson, NC: McFarland and Co., 1997.

Cooke, Chauncey H. "A Badger Boy in Blue: The Letters of Chauncey H. Cooke." *Wisconsin Magazine of History* 4, no. 1-4 (1920-21): 75-100, 208-17, 322-44, 431-56.

Cozzens, Peter. *General John Pope: A Life for the Nation.* Urbana: University of Illinois Press, 2005.

Cronon, William. *Changes in the Land: Indians, Colonists, and the Ecology of New England.* New York: Hill and Wang, 1983.

Dahlin, Curtis A. *Victims of the Dakota Uprising: Killed, Wounded, and Captured.* Roseville, MN: The Author, 2012.

DeCarlo, Peter. *Fort Snelling at Bdote: A Brief History.* St. Paul: MNHS Press, 2016.

Denial, Catherine J. *Making Marriage: Husbands, Wives, and the American State in Dakota and Ojibwe Country.* St. Paul: MNHS Press, 2013.

———. "Pelagie Faribault's Island: Property, Kinship, and the Meaning of Marriage in Dakota Country." *Minnesota History* 62, no. 2 (Summer 2010): 48-59.

Dyer, Frederick H. *A Compendium of the War of the Rebellion Compiled and Arranged from Official Records of the Federal and Confederate Armies.* 1908. Reprint, Bethesda, MD: University Publications of America, 1994.

Eastman, Mary. *Dahcotah; or, Life and Legends of the Sioux around Fort Snelling.* New York: J. Wiley, 1849.

Edmunds, R. David, ed. *American Indian Leaders: Studies in Diversity.* Lincoln: University of Nebraska Press, 1980.

Executive Documents of the State of Minnesota for 1862. St. Paul, MN: Wm. R. Marshall, 1863.

Folwell, William Watts. *A History of Minnesota.* 3 vols. St. Paul: MNHS Press, 1921-30.

Forsyth, Thomas. "Fort Snelling: Col. Leavenworth's Expedition to Establish It, in 1819." *Collections of the Minnesota Historical Society* 3 (1880).

Fremling, Calvin R. *Immortal River: The Upper Mississippi in Ancient and Modern Times.* Madison: University of Wisconsin Press, 2005.

Fridley, Russell W. "Fort Snelling: From Military Post to Historic Site." *Minnesota History* 35, no. 4 (Winter 1956): 178-92.

Gilman, Rhoda R. *Henry Hastings Sibley: Divided Heart*. St. Paul: MNHS Press, 2004.

———. "The Last Days of the Upper Mississippi Fur Trade." *Minnesota History* 42, no. 4 (Winter 1970): 122–40.

Gilman, Rhoda R., Carolyn Gilman, and Deborah L. Miller. *The Red River Trails: Oxcart Routes Between St. Paul and the Selkirk Settlement, 1820-1870*. St. Paul: MNHS Press, 1979.

Goodman, Nancy. *Paddlewheels on the Upper Mississippi, 1823-1854: How Steamboats Promoted Commerce and Settlement in the West*. Stillwater, MN: Washington County Historical Society, 2003.

Goodman, Nancy, and Robert Goodman. *Joseph R. Brown: Adventurer on the Minnesota Frontier, 1820-1849*. Rochester, MN: Lone Oak Press, 1996.

Green, William D. "Minnesota's Long Road to Black Suffrage, 1849-1868." *Minnesota History* 56, no. 2 (Summer 1998): 68–84.

———. *A Peculiar Imbalance: The Fall and Rise of Racial Equality in Minnesota, 1837-1869*. Minneapolis: University of Minnesota Press, 2015.

Haines, Michael R. "Estimated Life Tables for the United States, 1850-1900." Historical Paper 59. Cambridge, MA: National Bureau of Economic Research, 1994.

Hamilton, Holman. "Zachary Taylor and Minnesota." *Minnesota History* 30 (Summer 1949): 97–110.

Hansen, Marcus Lee. *Old Fort Snelling*. 1918. Reprint, Minneapolis: Ross and Haines, 1958.

Harper, Donald V. "The Minneapolis-St. Paul Metropolitan Airports Commission." *Minnesota Law Review* 55, no. 3 (1971): 363.

Harpole, Patricia C., and Mary D. Nagle, eds. *Minnesota Territorial Census, 1850*. St. Paul: MNHS, 1972.

Heitman, Francis B. *Historical Register and Dictionary of the United States Army*. Washington, DC: GPO, 1903.

Holbrook, Franklin F. *Minnesota in the Spanish-American War and the Philippine Insurrection*. St. Paul: Minnesota War Records Commission, 1923.

Holbrook, Franklin F., and Livia Appel. *Minnesota in the War with Germany*. St. Paul: MNHS, 1932.

Hunter, David. *Report of the Military Services of Gen. David Hunter, U.S.A., during the War of the Rebellion, Made to the US War Department, 1873*. New York: D. Van Nostrand, 1873.

Hyman, Colette A. "Survival at Crow Creek, 1863-1866." *Minnesota History* 61, no. 4 (Winter 2008-9): 148–61.

Johnson, Loren C. "Reconstructing Old Fort Snelling." *Minnesota History* 42, no. 3 (Fall 1970): 82–98.

Jones, Evan. *Citadel in the Wilderness: The Story of Fort Snelling and the Northwest Frontier*. Minneapolis: University of Minnesota Press, 2001.

Kane, Lucile M. *The Waterfall that Built a City: The Falls of St. Anthony in Minneapolis*. St. Paul: MNHS, 1966.

Larson, Peggy Rodina. "A New Look at the Elusive Inkpaduta." *Minnesota History* 48, no. 1 (Spring 1982): 24–35.

Lass, William E. "Histories of the U.S.-Dakota War of 1862." *Minnesota History* 63, no. 2 (Summer 2012): 44–57.

Le Duc, W. G. *Minnesota Yearbook for 1852*. St. Paul, Minnesota Territory: W. G. Le Duc, 1852.

Lehman, Christopher P. *Slavery's Reach: Southern Slaveholders in the North Star State*. St. Paul: MNHS Press, 2019.

Loehr, Rodney C. "Franklin Steele, Frontier Businessman." *Minnesota History* 27, no. 4 (Winter 1946): 309–18.

Long, Stephen H. *The Northern Expeditions of Stephen H. Long: The Journals of 1817 and 1823 and Related Documents*. Edited by Lucile M. Kane, June D. Holmquist, and Carolyn Gilman. St. Paul: MNHS Press, 1978.

Lovelace, Maude Hart. *Early Candlelight*. 1921. Reprint, St. Paul: MNHS Press, 1992.

Luecke, Barbara K., and John C. Luecke. *Snelling: Minnesota's First First Family*. Eagan, MN: Grenadier Publications, 1993.

Martin, Patrick R. "Forgotten Pioneer: Abraham Perry and the Story of His Flock." *Ramsey County History* 26, no. 3 (Fall 1991): 19-21.

Matsen, William E. "The Battle of Sugar Point: A Re-Examination." *Minnesota History* 50, no. 7 (Fall 1987): 269–75.

McMahon, Eileen M., and Theodore J. Karamanski. *North Woods River: The St. Croix River in Upper Midwest History*. Madison: University of Wisconsin Press, 2010.

Meyer, Roy W. *Everyone's Country Estate: A History of Minnesota's State Parks*. St. Paul: MNHS Press, 1991.

Millikan, William. "The Great Treasure of the Fort Snelling Prison Camp." *Minnesota History* 62, no. 1 (Spring 2010): 4-17.

Minnesota Board of Commissioners. *Minnesota in the Civil and Indian Wars, 1861-1865*. 1893. Reprint, St. Paul: MNHS Press, 2005.

Moe, Richard. *The Last Full Measure: The Life and Death of the First Minnesota Volunteers*. St. Paul: MNHS Press, 2001.

Monjeau-Marz, Corrine. *The Dakota Indian Internment at Fort Snelling, 1862-1864*. St. Paul, MN: Prairie Smoke Press, 2006.

Nankivell, John H. *The History of the Twenty-Fifth Regiment, United States Infantry. 1869-1926*. Fort Collins, CO: Old Army Press, 1972.

Newcombe, Barbara T. "'A Portion of the American People': The Sioux Sign a Treaty in Washington in 1858." *Minnesota History* 45, no. 3 (Fall 1976): 82-96.

Nute, Grace Lee. "Hudson's Bay Company Posts in the Minnesota Country." *Minnesota History* 22 (Fall 1941): 270–89.

Obst, Janis. "Abigail Snelling: Military Wife, Military Widow." *Minnesota History* 54, no. 3 (Fall 1994): 98–111.

Okrent, Daniel. *Last Call: The Rise and Fall of Prohibition*. New York: Scribner, 2010.

Olson, Russell L. *The Electric Railways of Minnesota*. St. Paul, MN: H. M. Smyth Co., 1976.

Orsi, Jared. *Citizen Explorer: The Life of Zebulon Pike*. Oxford: Oxford University Press, 2014.

Osman, Stephen. "Fort Is Sacred to the Memory of Those Who Served There." *Highland Villager*, November 16, 2019.

Osman, Stephen E. *Fort Snelling and the Civil War*. St. Paul, MN: Ramsey County Historical Society, 2017.

———. *Fort Snelling Then and Now: The World War II Years.* St. Paul, MN: Friends of Fort Snelling, 2011.

Parker, Donald Dean, ed. *The Recollections of Philander Prescott: Frontiersman of the Old Northwest, 1819-1862.* Lincoln: University of Nebraska Press, 1966.

Pond, Samuel. *The Dakota or Sioux in Minnesota as They Were in 1834.* 1908. Reprint, St. Paul: MNHS Press, 1986.

Prescott, Philander. "Autobiography and Reminiscences of Philander Prescott." *Collections of the Minnesota Historical Society* 6 (1894).

Pringle, Heather. "The First Americans: Mounting Evidence Prompts Researchers to Reconsider the Peopling of the New World." *Scientific American* (November 2011).

Prosser, Richard S. *Rails to the North Star.* Minneapolis: Dillon Press, 1966.

Prucha, Francis Paul. "Army Sutlers and the American Fur Company." *Minnesota History* 40, no. 1 (Spring 1966): 22-31.

———. "Fort Ripley: The Post and Military Reservation." *Minnesota History* 28, no. 3 (Fall 1947): 205-22.

———. *The Sword of the Republic: The United States Army on the Frontier, 1783-1846.* 1969. Reprint, Lincoln: University of Nebraska Press, 1987.

Ritchey, Charles James. "Claim Associations and Pioneer Democracy in Early Minnesota." *Minnesota History* 9, no. 2 (Summer 1928): 85-95.

Robinson, Charles M. III. *A Good Year to Die: The Story of the Great Sioux War.* Norman: University of Oklahoma Press, 1996.

Roddis, Louis H. "The Last Indian Uprising in the United States." *Minnesota History Bulletin* 3 (February 1920): 273-90.

Rorabaugh, W. J. *The Alcoholic Republic: An American Tradition.* New York: Oxford University Press, 1981.

Slovak, Marilyn L. "Smartest Horse in the U. S. Army: Whiskey of Fort Snelling." *Minnesota History* 61, no. 8 (Winter 2009-10): 336-45.

Smith, Hampton. "Minnesota on the Verge of Civil War." *Minnesota's Heritage* 4 (July 2011): 52-59.

Stewart, Richard W., ed. *American Military History.* Washington, DC: Center of Military History, US Army, 2005.

Summit Envirosolutions and Two Pines Resource Group. "The Cultural Meaning of Coldwater Spring: Final Ethnographic Resources Study." Prepared for the National Park Service, Mississippi National River and Recreation Area. June 2006. https://www.nps.gov/parkhistory/online_books/miss/coldwater_spring.pdf.

Swanberg, W. A. "Was the Secretary of War a Traitor?" *American Heritage* 14, no. 2 (February 1963).

Taliaferro, Lawrence. "Auto-biography of Major Lawrence Taliaferro: Written in 1864." *Minnesota Historical Society Collections* 6 (1894).

Taylor, David Vassar. *African Americans in Minnesota.* St. Paul: MNHS Press, 2002.

Tyson, Amy M. *The Wages of History: Emotional Labor on Public History's Front Lines.* Amherst: University of Massachusetts Press, 2013.

Upham, Warren. *Minnesota Geographic Names: Their Origins and Historical Significance.* St. Paul: MNHS, 1920.

———. *The Women and Children of Fort Saint Anthony, Later Named Fort Snelling.* New York: Magazine of History, 1915.

Van Cleve, Charlotte Ouisconsin. *"Three Score Years and Ten": Life-Long Memories of Fort Snelling, Minnesota, and Other Parts of the West.* Minneapolis: Harrison and Smith, 1888.

Van Cleve, Stewart. *Land of 10,000 Loves: A History of Queer Minnesota.* Minneapolis: University of Minnesota Press, 2012.

VanderVelde, Lea. *Mrs. Dred Scott: A Life on Slavery's Frontier.* New York: Oxford University Press, 2010.

Vane, Elizabeth A. P., and Sanders Marble. "Contributions of the U.S. Army Nurse Corps in World War I." *Soins: La revue de référence infirmière* (June 2014). https://e-anca.org/History/Topics-in-ANC-History/Contributions-of-the-US-Army-Nurse-Corps-in-WWI.

Viola, Herman J. *Little Bighorn Remembered: The Untold Indian Story of Custer's Last Stand.* New York: Times Books, 1999.

Watson, Catherine. "Fort Snelling as I Knew It." *Open Rivers: Rethinking Water, Place & Community* 7 (2017). https://editions.lib.umn.edu/openrivers/article/fort-snelling-as-i-knew-it/.

Weigley, Russell F. *History of the United States Army.* Bloomington: Indiana University Press, 1984.

Wells, Spencer. *The Journey of Man: A Genetic Odyssey.* 2002. Reprint, Princeton, NJ: Princeton University Press, 2017.

Westerman, Gwen, and Bruce White. *Mni Sota Makoce: The Land of the Dakota.* St. Paul: MNHS Press, 2012.

White, Bruce. "Indian Visits: Stereotypes of Minnesota's Native People." *Minnesota History* 53, no. 3 (Fall 1992): 99-111.

White, Helen M. *The Tale of a Comet and Other Stories.* St. Paul: MNHS Press, 1984.

White, Helen M., and Bruce M. White. *Fort Snelling in 1838: An Ethnographic and Historical Study.* St. Paul, MN: Turnstone Research, 1998.

Williams, J. Fletcher. *A History of the City of St. Paul to 1875.* St. Paul: MNHS Press, 1982.

Wills, Jocelyn. *Boosters, Hustlers, and Speculators: Entrepreneurial Culture and the Rise of Minneapolis and St. Paul, 1849-1883.* St. Paul: MNHS Press, 2005.

Wingerd, Mary Lethert. *Claiming the City: Politics, Faith, and the Power of Place in St. Paul.* Ithaca, NY: Cornell University Press, 2003.

———. *North Country: The Making of Minnesota.* Minneapolis: University of Minnesota Press, 2010.

Woodall, Allen E. "William Joseph Snelling and the Early Northwest." *Minnesota History* 10, no. 4 (Winter 1929): 367-85.

Index

177; supply problems, 71-72; sutlers, 53, 104, 246n15, 247n33 (chapter 4); "total army" idea, 199; volunteers to augment, 140; under Washington, 17-18. *See also specific forts; specific units*

US military during World War I: conscription during, 185, 199; medical treatment, 186, *188,* 188-95, *189, 190-91, 192, 194;* officers for, 185-86; United States Expeditionary Force casualties, 186; and value of "citizen soldiers," 199; weapon technologies, 186

US Supreme Court, 56

Van Cleve, Charlotte, 110
Van Sant, Samuel R., 197
Veterans Administration (VA), 195, 225, 230
Villa, Pancho, 183
Vose, Josiah: background, 244n11 (chapter 3); in command of Fort Snelling, 83; construction at St. Peter's, 30-31; death, 244n11 (chapter 3); and Josiah Snelling, 65

Wabasha: background, 14; and Forsyth's assurances, 29; and treaties, 122, 124, 152
War of 1812, 23
Warren, Lyman, 93
Washburn, Cadwallader, 128
Washington, George, 17-18
Wayne, Anthony, 18
Waziyatawiŋ (Angela Cavender Wilson), 237
Weckerling, John, 221
Whipple, Henry, 151-52
whiskey. *See* alcohol
Whiskey (horse), 208, *208*
White, John, 35
Whitehorne, Samuel, 110

Wilkinson, James, 19
Wilkinson, Melville, 178
Williams, J. Fletcher, 126
Willmar Tribune, 194-95
Wilson, Henry, 115
Wilson, Woodrow, 185
Wingerd, Mary: alcohol and fur trade, 15; race of inhabitants in area surrounding Fort St. Anthony, 51; "taming" of frontier myth and interactions with Native Americans, 11-12
Wita Taŋka (Big Island), 1
women: Army Nurse Corps, *190-91,* 191-92; attacks on Ho-Chunk, 77; code of, in nineteenth century, 52; European American, and men's marriages to Native women, 89; laundresses at each post, 42; wives of enlisted men, 53; wives of officers, 52-53. *See also* Dakota women
Wood, Leonard, 213
Works Progress Administration (WPA), 212, 215, 227
World War I, 198. *See also* US military during World War I
World War II: before US entry, 213, 215; expansion of Fort Snelling after outbreak of World War II in Europe, 215, *216;* Fort Snelling as recruitment and induction center during, 215, *216,* 217, *217, 218,* 219-20, *220,* 255n2 (chapter 13); Fort Snelling as separation center after, 222-23; internment of Japanese Americans, 221-22; soldiers stationed at Fort Snelling during, 255n9; specialized military training units at Fort Snelling, 220-21, *221, 222;* troops at Fort Snelling during, 255n9; US entry, 217

Yellowstone Expedition, 26-27

Confluence: A History of Fort Snelling

was designed and set in type by Judy Gilats in St. Paul, Minnesota.

The text face is Cardea and the display faces are

IM FELL DW Pica Pro and Proxima Nova.